The Hybrid Mind

The Hybrid Mind: The Human-AI Convergence explores the transformative potential of humans and AI systems collaborating to revolutionize risk management, oversight, and adaptive governance. Introducing Agentic AI systems capable of autonomous decision-making while preserving human authority, the book outlines a hybrid inference architecture that combines large and small language models, machine learning, and probabilistic reasoning.

Structured around the OODAF (Observe, Orient, Decide, Act, Feedback) loop, the book offers accessible explanations of intelligent system behavior and its implications. Use cases in finance, healthcare, cybersecurity, and SaaS environments demonstrate how hybrid intelligence enables scalable yet accountable operations, providing practical insights for the AI era.

It introduces original frameworks, including *Human ETHOS*, the *Experimentum Crucis Technology Matrix* (ECTM), and the *Hybrid Audit Model* (HAM), that reimagine ethical design, system evaluation, and real-time auditability. Chapters explore topics such as feedback architecture, ethical interruption, decision traceability, and context framing, critical elements for governing AI with integrity.

Designed for business leaders, technologists, and policymakers, *The Hybrid Mind* serves as both a strategic guide and a moral blueprint for the AI era. It equips readers to lead in a world where intelligence is shared, automation is consequential, and responsibility must be architected, not assumed.

Joseph X Ng is a seasoned executive with more than two decades of experience spanning finance, technology, academia, and the public sector. Known for his strategic leadership in technology innovation and operational excellence, he has led transformative initiatives at global enterprises. As former Head of the Global AI Community of Practice at a major financial institution, Joseph spearheaded the integration of AI and machine learning into enterprise systems, fostering scalable, responsible innovation. He currently mentors university graduate students in IT strategy and project execution. His writing bridges the worlds of executive leadership, advanced technology, and human-centered learning. Through his books and academic work, Joseph invites readers to explore how visionary leadership and emerging technologies can redefine industries and empower future leaders.

The Hybrid Mind
The Human-AI Convergence

Joseph X Ng

CRC Press
Taylor & Francis Group
CHAPMAN & HALL

Designed cover image: Shutterstock

First edition published 2026
by CRC Press
2385 NW Executive Center Drive, Suite 320, Boca Raton FL 33431

and by CRC Press
4 Park Square, Milton Park, Abingdon, Oxon, OX14 4RN

CRC Press is an imprint of Taylor & Francis Group, LLC

© 2026 Joseph X Ng

ISBN: 978-1-041-09077-9 (hbk)
ISBN: 978-1-041-09076-2 (pbk)
ISBN: 978-1-003-64833-8 (ebk)

DOI: 10.1201/9781003648338

Typeset in Times
by Newgen Publishing UK

Dedication

Dedicated to Tasha for her patience, inspiration, and support.

Contents

PART IV   Human-Centered Collaboration

PART V   Forward Humanity

Preface

NOT A THANG UNTIL IT IS

We rarely recognize a threshold when it arrives. It doesn't announce itself with certainty or clarity. One moment, a tool is on the fringe, an experiment on the margins of industry or research. The next, it's essential. Invisible. Embedded in our routines, in our language, in our very sense of normal. That shift, that almost imperceptible leap from irrelevance to inevitability, is what we call the thang moment. The tipping point is when the future stops waiting and starts unfolding. It isn't just a catchy phrase. "Not a thang until it is" is a worldview. This is a reminder that transformation doesn't always knock; sometimes, it just appears in your inbox or answers your question before you even ask.[1]

> "Not a thang until it becomes a thang™" isn't just a saying – it's a mindset. It captures the spark when an idea catches fire, when something overlooked becomes undeniable. From startups finding product-market fit to cultural shifts that rewrite entire industries, the thang moment is the tipping point, the leap from obscurity to inevitability. Life moves fast, and the magic often lives in the unexpected. So, stay ready. Stay innovative. Because when it hits, you'll know – it's officially a thang.[1]

This book begins at that threshold. Where serendipity meets system. Where intuition wrestles with automation. Where the human self, the core of judgment, creativity, and meaning, is asked to collaborate with something not quite human and not quite neutral. In a world awash in synthetic output, algorithmic decisions, and auto-piloted workflows, many are feeling a quiet unease, that sense of being adjacent to brilliance yet oddly displaced by it. You're in the room. You're doing the work. But the machine is finishing your sentence, anticipating your next click, and slowly pulling the center of gravity away from you. We refer to this dissonance as Synthetic Syndrome™, and it is becoming the emotional hallmark of the AI era.[2]

This is not about fearmongering. It's about naming what we feel. Because feelings, though often undervalued in technical circles, are data, too, signals that something meaningful is shifting. The Hybrid Mind™ recognizes this. It frames augmentation not just as a performance enhancement but as an identity recalibration.[3] It asks not only what AI can do but what remains uniquely ours to preserve. And at the heart of that inquiry is the concept this book formalizes: Human ETHOS™.[4]

Human ETHOS is not sentimentality. It is the human perspective made formal – a uniquely situated, emotionally aware, ethically bounded lens through which people interpret information, weigh consequences, and act with intention.[3] It resists reduction. It cannot be fully encoded. And yet, it is indispensable. ETHOS is what keeps hybrid collaboration human. In a time when algorithms simulate empathy and models mimic voice and tone, Human ETHOS is our internal compass that grounds speed in wisdom and ensures that automation does not eclipse intention. It preserves the moral texture of decision-making by recognizing that every action leaves a trace. This is the principle of karma: that choices in systems have consequences, both visible and invisible. To preserve the humanity of human-AI collaboration, we must maintain ETHOS, not as a safeguard, but as a moral rhythm woven into the loop itself.[5] It ensures that speed does not eclipse wisdom, and automation does not override intention.[3]

The Hybrid Mind, then, is both a mindset and a manifesto. It is how we orient ourselves when the map is being redrawn in real time. It is how we operate when certainty disappears, but responsibility remains.[6] This framework doesn't ask for technical perfection. It asks for adaptive clarity – the ability to know when to defer, when to intervene, and when to slow the loop, even when every system around you is urging acceleration.[7]

We are not building this book on the premise that AI will replace us. We make it the truth that AI will reshape us – and that we have a say in how.[8] This is a book of readiness. Of rhythm. Of recovery of self amid synthetic surges. Each chapter speaks to a shift already underway – from tools to teammates, from tasks to orchestration, from prediction to meaning. But it all begins here – with the decision to be present at the edge of change, not passive in its wake.[9]

There is always a moment before the moment. A flash where the news is still deniable. Still mockable. Still, "not a thang." That is when it's most vital to be ready. Because by the time everyone else notices, it's already too late to lead.[1]

So, stand here with your Human ETHOS intact. Let it guide you through the noise, the novelty, and the nuance because leadership in the era of AI begins not with code or compliance, but with clarity. A clarity that says: I see what is coming. I know what I bring. I am not being replaced, I am becoming, a Hybrid Mind.[10]

REFERENCES

1. Joseph X. Ng, *The Hybrid Mind: The Human-AI Convergence* (2025).
   Ng introduces the phrase "Not a thang until it becomes a thang" to frame the emotional and technological threshold when experimental tools quietly cross into cultural ubiquity, setting the thematic tone for the book.
2. Anna Kelly, "AI and the Erosion of Authorship," *Financial Times*, March 2025.
   Kelly's exploration of authorial displacement in the age of AI lends cultural relevance to the Preface's framing of "Synthetic Syndrome" as a psychological marker of creative erosion.
3. Fei-Fei Li, "Human-Centered AI: Building Trust with Empathy and Ethics," Stanford HAI Lecture, 2021.
   Li's vision of empathetic AI design supports the Preface's call to preserve interpretability and the human layer in intelligent systems, even as automation accelerates.
4. Joseph X. Ng, *EXAI Manifesto: Ethical eXcellence in Artificial Intelligence (EXAI) Manifesto* (Humainority.org, 2024), https://Humainority.org/exai-manifesto/.
   This manifesto defines the Human ETHOS as ethically bounded, emotionally aware, and essential to responsible AI augmentation. The Preface draws on these ideas to argue that feelings are valid data in a synthetic world.

5. Joseph X. Ng, Mastering IT Innovation: 90-Day Proof of Value Blueprint (New York: SAGE Ensights, 2024).
This work introduces the concept of innovation inflection points through the lens of proof-of-value cycles, a foundation for understanding the "thang moment" described in the Preface.

6. Joseph X. Ng and Andres Fortino, Strategic Innovation Governance: Identifying Crucial Use Cases via Experimentum Crucis Technology Matrix (ECTM) (New York: SAGE Ensights, 2025).
ECTM argues for decision frameworks that prioritize readiness and context over novelty. This reinforces the Preface's call to recognize transformation before it becomes inevitable.

7. Satya Nadella, *Hit Refresh: The Quest to Rediscover Microsoft's Soul and Imagine a Better Future for Everyone* (New York: HarperBusiness, 2017).
Nadella distinguishes between AI comprehension and human resonance, a point quoted in the EXAI Manifesto and echoed in the Preface's emphasis on emotional interpretability.

8. Joseph X. Ng, *Data Insights: Core Principles of Statistical Analysis for Storytelling* (New York: SAGE Ensights, 2024), https://a.co/d/edWAFvL.
Ng's insistence on human-led storytelling through data complements the Preface's positioning of the Human ETHOS as an irreplaceable interpretive layer in augmented systems.

9. Sherry Turkle, Reclaiming Conversation: The Power of Talk in a Digital Age (New York: Penguin Press, 2015).
Turkle's analysis of digital detachment and the diminishing space for real human presence resonates with the Preface's meditation on being "adjacent to brilliance yet displaced by it."

10. West, S., and Goff, B., "Agentic AI and the Design of Ethical Hesitation," *Design Ethics Quarterly*, 2024.
The Preface's closing emphasis on knowing when to slow the loop draws from this article's concept of "ethical hesitation," advocating for hybrid discernment over automated default.

AI Usage Disclaimer

Select generative AI tools (OpenAI's ChatGPT-4 and DALL-E 3) were used during the development of this book in a strictly assistive capacity. These tools supported early-stage ideation, language refinement, and figure concept drafting. All substantive content – including the book's arguments, original frameworks, analytical narrative, and conclusions – was conceived, written, and finalized by the author. No generative AI tool was credited with authorship, nor did any AI system contribute autonomously to the book's core intellectual content.

Image generation tools, where used, have been disclosed in the captions in accordance with the publisher's policy. All AI-assisted images were created under human guidance and reviewed for accuracy, ethics, and relevance.

The author assumes full responsibility for the integrity and originality of the content presented in this work.

Part I

Becoming the Hybrid Mind

FROM OPERATOR TO ORCHESTRATOR, WHERE LEADERSHIP MEETS AUGMENTATION

The idea of the Hybrid Mind emerges at a time when the line between human cognition and machine augmentation is no longer theoretical, it is practical, lived, and accelerating. For decades, humans have developed systems to extend their capabilities, from calculators and spreadsheets to machine learning and generative models. However, augmentation today goes beyond extension. It shifts the nature of agency. This part begins with the recognition that the future of leadership will be defined not by how well one executes tasks but by how well one orchestrates systems of intelligence, across contexts, across machines, and ethical boundaries.

The concept of the Hybrid Mind challenges the notion of replacement. Instead of viewing automation as a zero-sum game, where the machine triumphs and the human is left behind, the Ethical eXcellence in Artificial Intelligence (EXAI) Manifesto™ proposes a different perspective. It reframes the relationship as one of augmentation: AI is not a competitor but a partner. It does not supplant judgment; rather, it enhances the ability to make informed decisions in complex situations.[1] This marks a significant cognitive and cultural shift. In this model, the Hybrid Mind is not engaged in linear workflows but in orchestrating multidimensional systems, curating inputs, managing feedback loops, and aligning decisions with context.

Orchestration, in the context of data-driven organizations, is not about traditional control. It is a form of leadership that requires discernment, sensemaking,

DOI: 10.1201/9781003648338-1

and ethical arbitration. Today's leaders are not just managing people but also the interplay between algorithms, metrics, and meaning. The shift from operator to orchestrator is characterized by the ability to set parameters, define what matters, and intervene when systems deviate from their intended purpose. For instance, the orchestrator in a data-driven organization does not code models line by line; instead, they define the architecture of evaluation, choose which data to analyze, and decide when a human must remain in the loop.[2] This is not a technical function. It is a moral one.

The most valuable contribution of AI is not its ability to answer questions but its capacity to generate context. The system uncovers hidden patterns, maps the decision space, and presents alternative futures. It is a context engine, not a deterministic solution provider. In this sense, the power of the Hybrid Mind is not in knowing more, it is in asking better questions. By surfacing novel connections, Hybrid Intelligence not only accelerates cognition but also reshapes what counts as knowledge. The Experimentum Crucis Technology Matrix (ECTM)™, introduced as a model for strategic use-case selection, reinforces this idea by asking leaders to evaluate not only feasibility but also the ethical, cultural, and systemic consequences of adoption.[3]

The technical foundation of this mindset is modularity. Multi-modular inference replaces the monolithic intelligence model with one that draws from small, specialized components: symbolic reasoning, probabilistic models, retrieval engines, and generative layers. This architecture reflects the human mind more than any single algorithm does. Just as people do not think in a single mode, logic, intuition, memory, and emotion operate in tandem, hybrid systems benefit from orchestrating different AI personalities tuned to specific domains. This is not just a matter of efficiency. It is a philosophical stance about the diversity of thought and the importance of contextual relevance.[4]

However, even with this promise, the paradox of agency lingers. As we embrace collaboration with increasingly autonomous systems, the question arises: when do we step back, and when must we step in? This is the paradox between control and collaboration. The more capable a system becomes, the more tempting it is to let go of control. However, relinquishing too much introduces risk, of drift, opacity, and ethical blind spots. The Hybrid Mind must be capable of ethical hesitation, a term derived from discussions of Agentic AI, which emphasizes the design of systems that allow for pause, reconsideration, and human override in moments of moral complexity. It is in this hesitation, not in speed, that leadership shows up.

What becomes clear is that intelligence is no longer what happens inside a person or a machine. It is what happens between them. The Hybrid Mind is the relational layer, the active presence that decides how systems interact, how meaning is made, and how values are preserved. To lead in this new era is to hold both the machine's confidence and the human's doubt, not as contradictions, but as co-necessary parts of adaptive intelligence. It is not about knowing the correct answer. It is about creating the conditions where the proper process can emerge.

REFERENCES

1. Joseph X. Ng, *EXAI Manifesto: Ethical eXcellence in Artificial Intelligence (EXAI) Manifesto* (New York: Humainority.org, 2025), https://Humainority.org/exai-manifesto/.
    This manifesto lays the philosophical foundation for augmentation over automation, reframing AI as a partner in human decision-making. It defines the ethical responsibilities of the Hybrid Mind, including the right to explanation, refusal, and moral override in hybrid systems.
2. Joseph X. Ng, *Mastering IT Innovation: 90-Day Proof of Value Blueprint* (New York: SAGE Ensights LLC, 2024), https://a.co/d/07ZO4zW.
    Ng outlines how innovation leadership moves beyond managing pilots to demonstrating lasting value. This supports the chapter's emphasis on the orchestrator role, where leaders are expected to direct dynamic systems, govern AI ensembles, and ensure use cases align with enterprise purpose.
3. Joseph X. Ng and Andres Fortino, Strategic Innovation Governance: Identifying Crucial Use Cases via Experimentum Crucis Technology Matrix (ECTM) (New York: Humainority, 2025).
    This white paper introduces the ECTM framework, which the chapter applies as a mechanism for strategic discernment. Rather than blindly scaling experimentation, the Hybrid Mind evaluates whether a use case is not only technologically feasible but ethically and operationally transformative.
4. Joseph X. Ng, *Data Insights: Core Principles of Statistical Analysis for Storytelling* (New York: SAGE Ensights LLC, 2024), https://a.co/d/edWAFvL.
    This book emphasizes that data gains meaning only when interpreted through human-centered narratives. The Hybrid Mind draws on this perspective to argue that AI's true value lies not in answers, but in generating richer contexts for human decision-making.

1 The Hybrid Mind
A New Era of Decision-Making

At the intersection of thought and code, a new form of intelligence is emerging – the Hybrid Mind. Neither fully human nor wholly machine, it thrives in the space where intuition meets algorithm. In a world where AI is no longer optional but omnipresent, the real question is not whether we use it, but how we make the collaboration count. The quality of our collaboration, as tech professionals, AI researchers, futurists, and innovation enthusiasts, is crucial in shaping the future of AI. The Hybrid Mind is not a metaphor. It is a decision framework, a leadership mindset, and a systems design philosophy that asks a simple question: how do we stay human in systems that are becoming smarter by the hour? The potential of the Hybrid Mind to revolutionize the future of work is immense, offering a new perspective and optimism.

We begin by defining Hybrid Intelligence as more than augmentation; it is a strategic fusion. Humans bring judgment, ethics, and intuition. Machines bring speed, scale, and inference. Together, they shape decisions that neither can reach alone. From there, the chapter explores how leadership roles evolve from managing tasks to orchestrating complex, modular systems. In this shift, AI becomes less about execution and context, surfacing unseen patterns, anticipating consequences, and reframing what it means to understand.

The chapter also introduces Multi-Modular Inference as the architecture of hybrid thinking. Rather than relying on a single, all-knowing model, effective AI systems now operate as collaborative networks of specialized agents, each performing distinct functions that are coordinated by human intent. Finally, we confront the paradox at the center of all hybrid collaboration: as we gain speed and insight through machines, we also risk losing agency. The solution lies not in control but in conscious collaboration, where design, ethics, and emotional intelligence guide our partnerships with AI.

This chapter sets the tone for the book by reframing the concept of decision-making itself. It invites readers to move beyond debates about automation and toward a richer conversation about co-creation, responsibility, and relevance in an intelligent world.[1]

DOI: 10.1201/9781003648338-2

1.1   THE RISE OF HYBRID INTELLIGENCE

There comes a quiet shift in every transformative age, a moment when the old frameworks of thought no longer suffice. We are in one of those moments now. Not because machines have become sentient but because they have become embedded, woven into the rhythm of our decisions, designs, and everyday sense of possibility. What emerges from this entanglement is not a contest between humans and machines but a collaboration, a convergence. Hybrid Intelligence is the name we give this union: the strategic integration of human intuition and ethical reasoning with machine-driven computation, pattern recognition, and scalability. It is a system that includes us all.

Unlike narrow artificial intelligence, which operates in predefined domains, Hybrid Intelligence is not constrained by a single function or form. It thrives in ambiguity. It is less about machines replacing humans and more about redefining roles, humans as curators of context, sense makers of ambiguity, and arbiters of meaning. In this framework, the machine is not merely a tool that executes; it is a context engine, surfacing options, identifying gaps, and simulating outcomes. Nevertheless, the final synthesis, the integration of what matters and why, still belongs to us.

Consider a financial risk officer who uses machine learning to scan millions of transactions in real-time. The system surfaces anomalies, predictive indicators, and correlations the human eye would miss. However, when it comes to halting a trade or alerting regulators, the final decision is layered with human judgment that weighs reputational risk, institutional precedent, and regulatory nuance. The machine offers scope. The human brings shape.

This dynamic is not hypothetical. Researchers at the Wharton School refer to it as "Hybrid Intelligence," defining it as "the computational strengths of AI fused with the holistic comprehension of human cognition," a model designed for sustainability, creativity, and trust in decision-making.[2] Rather than viewing AI as a replacement, the model reframes it as a partner that augments our vision without displacing our value.

However, with that vision comes tension. As previously explored in my newsletter Synthetic Syndrome, the very success of AI systems in creative domains – writing, illustrating, and composing – can generate a kind of emotional dissonance in professionals who once took pride in their uniqueness. Not just coders or artists feel this; even senior leaders have expressed discomfort when AI systems outperform their strategic intuitions.[3] This erosion of confidence is not born from failure but from proximity to systems that feel uncannily capable. The imposter feeling here is quiet, almost ambient, a byproduct of accelerated augmentation. We must acknowledge it if we are to navigate it.

The Hybrid Mind, then, is not merely a means of working. It is a way to see. It demands that we shift from thinking about productivity in terms of linear tasks to thinking about intelligence in distributed orchestration. The question is no longer "What can AI do?" but "What do we want it to mean?" Reframing our relationship to intelligent systems, from toolmakers to symbiotic designers, we reframe ourselves. We become not users but stewards. Not operators, but orchestrators. Moreover, in this role, we find reassurance in the adaptability of the Hybrid Mind.

As we continue this chapter, remember that Hybrid Intelligence is not a distant future but a present unfolding unevenly. It is evident in the decisions of healthcare teams utilizing AI diagnostics, in the choices of educators designing with adaptive learning tools, and in the strategic posture of executives evaluating use cases through modular AI governance frameworks, such as the ECTM.[1] It is a mindset, not just a method. In that mindset, the human core is not lost; it becomes new and vital.

1.2   FROM TASK EXECUTION TO SYSTEM ORCHESTRATION

The earliest use cases for AI were built around specific tasks: predicting a value, classifying an image, and answering a question. These were narrow, transactional roles – akin to automating a single step in a process. However, as systems evolved, so did the expectations. The modern challenge is not how well a system can perform a single task but how well it can coordinate multiple tasks. In that shift, we begin to see the actual value of hybrid intelligence, not as a way to automate work but to orchestrate complexity.

Task execution assumes clarity. There is a beginning, a middle, and an end. You feed a model data; it returns a prediction. However, real-world decisions – particularly those involving strategy, ethics, and risk – rarely conform to that model. They are messy. Contextual. Interdependent. In these spaces, a single model cannot govern the whole process. What is needed is an architecture of orchestration – a way to coordinate multiple inputs, perspectives, and probabilities into coherent action. This is the role of the Hybrid Mind.

This orchestration was framed as a technical and a leadership function in the Strategic Innovation Governance paper. Technical in the sense that systems must interoperate, passing signals across APIs, logging outputs for audit, and surfacing decision branches transparently. However, it is also a leadership act that requires prioritizing competing outcomes and weighing trade-offs beyond mere optimization.[1] A chief innovation officer is not just greenlighting AI projects, they are curating a portfolio of decisions that affect ethics, operations, and public trust.

Consider a case in healthcare. A hospital system might use one model to estimate readmission risk, another to generate patient summaries, and a third to recommend care pathways. Each system performs well individually. However, they were not designed to talk to each other. The burden of integration falls on the clinician, who must synthesize these inputs under time pressure. It is not augmentation; it is fragmentation. Orchestration means designing the system not just to give outputs but to anticipate how those outputs inform one another. The Hybrid Mind fills the gaps between tools, creating a layer of logic and continuity that machines alone cannot hold.

It also reframes the role of leadership in hybrid environments. Instead of managing tasks, leaders shape environments. They decide which systems interact, how data flows, and where human checkpoints are essential. This is echoed in Boyd's OODA loop, which we explore in the next chapter, as the need to slow down and reorient becomes increasingly critical as automation speeds up.

The Hybrid Mind Outline captures this transition as a shift from operator to orchestrator, a progression that mirrors the evolving definition of intelligence itself. Intelligence is no longer about knowing the answer but knowing which systems to ask, how to translate outputs, and when to override. The modern leader is part conductor, part ethicist, and part systems architect.

Orchestration also introduces new responsibilities. When decisions are distributed across people and machines, accountability becomes diffused. Who is responsible for a failure when a recommendation engine, a sentiment analysis model, and a human analyst all play a role? This is why frameworks like ECTM matter.[1] They provide scaffolding for orchestrated environments, mapping out where human decision points must be preserved and where delegation to machines is safe and justifiable.

Moving from task to orchestration is a technical upgrade and a cognitive leap. It asks us to think in layers, flows, and contingencies. It demands an ability to balance control with collaboration and to remain calm in the presence of systems that often seem smarter than we are. However, the orchestration layer is still human. Moreover, it is there, in the in-between, in the quiet integration of the many, that the Hybrid Mind does its real work.

1.3 AI AS CONTEXT ENGINE, NOT JUST ANSWER ENGINE

Most people still approach artificial intelligence expecting an answer: Ask the machine, and you will get a response. However, the true power of AI, and its value to decision-makers, lies not in the answer but in the context it constructs. Context is what gives answers meaning. It explains relevance, traces assumptions, and reveals gaps. In hybrid decision environments, context is everything.

When an AI system recommends a course of action, the output is not the end of thinking; it is the beginning of alignment, alignment across teams, functions, risks, and regulations. Consider an AI platform used in logistics. A model may recommend rerouting freight to avoid weather delays. However, the implications stretch further: How will this affect warehousing? What happens to perishable inventory? Are there regional policies affected by this change? A traditional model focuses on efficiency. A context engine surfaces dependencies.

AI can generate context, but only human perspective gives it meaning.[3] It is our lived sense of what matters, why it matters, and to whom that transforms data into relevance. This gap between algorithmic context and lived comprehension is the new frontier. Without human framing, context is computational scaffolding – present but hollow. With it, it becomes the backbone of ethical, responsible decision-making.

This shift is becoming critical as AI is deployed in high-impact domains. In the financial services industry, AI is utilized to identify anomalies in transactions. However, flagging a transaction is not enough. The investigator must know why it was flagged, what rules were triggered, and how the behavior compares to past cases. The output alone is not trustworthy unless it is traceable. As the Strategic Innovation Governance paper outlines, AI serves as a context engine, reframing the relationship between speed and trust.[1] The goal is not to answer faster but to reason more deeply, especially in environments where unintended consequences compound quietly over time.

This perspective also transforms how leaders and regulators must think. A recommendation made by a machine cannot be accepted blindly. It must be situated. Is it aligned with compliance policies? Does it uphold ethical commitments? How does it intersect with workforce impacts, ESG (Environmental, Social, and Governance) criteria, or long-term strategy? These questions point toward explainability not as a technical bonus but as a strategic necessity. It is the only way humans stay meaningfully in the loop.[4]

Emerging literature echoes this idea. In their analysis of intelligent development environments, Data.World calls context engines "a design pattern for AI that supports explainability through traceable, layered logic."[2] Rather than returning predictions alone, a context engine reconstructs the setting in which that prediction is made. This turns AI into a collaborator, not a shortcut.

Hybrid Intelligence, then, must be built around context loops. Every output should provoke a human question: What am I not seeing? What does this assume? What else matters here? The orchestration layer discussed in the previous section becomes the space where context is gathered, aligned, and ultimately weighed against competing goals. It is where the human role becomes irreplaceable.

The EXAI Manifesto's concept of lifecycle accountability depends on this idea. Tracing a model's decision is not enough; we must also trace our decision-making frameworks, how we weigh risk, prioritize fairness, and respond to ambiguity. AI can support this, but only if it builds context around data, not just predictions from it.[3]

Context prevents AI from becoming a black box of synthetic reasoning. Without it, even accurate systems become untrustworthy. However, with it, the Hybrid Mind has the raw material to make sense, steer wisely, and pause with purpose. At the core of all strategic leadership is not the pursuit of the perfect answer, it is the relentless search for a deeper understanding.

1.4   MULTI-MODULAR INFERENCE: THE ARCHITECTURE OF HYBRID THINKING

If the Hybrid Mind is a way of thinking, its architecture is modular. The dominant myth in artificial intelligence still leans toward monoliths, the idea that a single, massive model can handle every problem. This approach is brittle, opaque, and insufficient for complex decision environments. Real-world systems require various types of reasoning, including symbolic, probabilistic, contextual, and task-specific reasoning. The only way to scale that diversity is to build intelligence not as a block but as a network. This is where multi-modular inference comes in.

Modular systems combine different AI components with a specialized function into coordinated frameworks that reflect how organizations operate. A legal firm might deploy a natural language model to summarize briefs, a rule-based system to validate against policy, and a statistical model to estimate case success rates. In isolation, each of these modules excels at a narrow task. Nevertheless, their true power is realized when orchestrated, sequenced, filtered, and layered to address more nuanced human goals. This form of distributed reasoning is not only more efficient but also more comprehensible, auditable, and adaptable.

In the "AI in Action" file, multi-modular inference is described as a response to the limitations of large monolithic models, such as GPT-4, which, while powerful, often lack logical consistency, explainability, and domain specificity.[5] Organizations can create composite intelligence that reflects real-world trade-offs and domain constraints by integrating smaller, purpose-built language models with rule engines, machine learning classifiers, and retrieval systems. It is less about building one model to rule them all and more about building many models that understand each other.

This approach mirrors the evolution of software architecture. Just as microservices have replaced the monolithic application model in cloud computing, modular AI is replacing the one-size-fits-all promise of LLMs. A customer support system might use an LLM to interpret user queries, route compliance questions to a rule engine, and direct emotional sentiment analysis to a fine-tuned classifier. The result is not just higher accuracy but better oversight. Modularity enables the isolation and repair of the failure point when something goes wrong. When something goes right, it can be attributed to the correct model and improved accordingly.

The implications of this shift are significant. First, it changes the economics of AI deployment. Companies can adopt modular components that address specific needs rather than investing in massive models with unclear ROI. This "use case first" mentality echoes the Experimentum Crucis Technology Matrix (ECTM), which frames modularity as both an architectural and investment advantage.[1] A modular approach enables incremental scaling, building capability over time rather than relying on a single model.

Second, modular inference enables broader participation in AI design. A domain expert, without needing to understand deep learning, can contribute to a rule set or design an SLM (Small Language Model) tuned for legal compliance. A systems architect can define orchestration patterns. A C-suite executive can direct which outputs are tied to which KPIs (Key Performance Indicators). This democratization of design enables organizations to align their AI with business goals rather than just engineering possibilities.

Even trust improves. Multi-modular systems naturally resist the "black box" trap. Because each module has a defined role and well-defined input-output behavior, it is easier to explain the system's behavior. In finance, healthcare, law, and education, this is not just a technical feature but a regulatory requirement. Modularity is the only viable design in environments where audibility and accountability are non-negotiable.

Imagine a hybrid AI used in university admissions to see this in action. A large language model interprets personal statements. A rules engine enforces eligibility requirements. A statistical model evaluates academic risk. Together, these systems create a multidimensional profile of each applicant, not as a number but as a decision context. It is Hybrid Thinking. Furthermore, it is built, not trained.

1.5   THE PARADOX OF CONTROL VS. COLLABORATION

An unsettling irony is at the heart of human-AI collaboration: the more intelligent the system becomes, the more we must let go to use it effectively. However, in doing so, we risk losing sight of who makes the decisions. It is the paradox of control versus collaboration. In hybrid systems, the human is still the decision-maker in theory. However, the machine has increasingly become the first responder – surfacing options, generating solutions, and even nudging action – and the line between recommendation and automation blurs.

At first glance, this seems efficient. Why question a system that gets it right most of the time? Nevertheless, that is precisely the problem. Sound systems do not eliminate oversight – they make it easier to forget. The smoother the interface, the less friction there is to pause. Moreover, that pause is where meaning, ethics, and leadership often reside.

Agentic AI exemplifies this shift. As introduced in later chapters, agentic systems do not merely respond to inputs; they initiate actions, learn from feedback, and maintain persistent goals. When a system generates content and decides when and how to intervene, the human's role moves from pilot to co-pilot – or worse, to passenger. The Hybrid Mind must actively reclaim its position – not through override but through the intentional orchestration of collaboration boundaries.[6]

It does not mean pulling the plug. It means designing for ethical hesitation. A well-designed hybrid system knows when to slow down, when to request confirmation, and when to acknowledge uncertainty. The EXAI Manifesto refers to this lifecycle accountability as the idea that human agency must be preserved not only at the point of deployment but throughout the entire operational lifespan of the system.[7] Collaboration without accountability is not augmentation, it is abdication.

In practice, this plays out in small but significant ways. Consider a generative AI model an HR team uses to draft performance reviews. The system can access employee metrics, sentiment analysis, and peer feedback. It can write a review in seconds. However, what if it misreads tone? What if it encodes bias from previous reviews? Who is accountable for the result if the manager accepts the output? The manager may still sign the form, but the real decision that shaped meaning was already made upstream.

In this context, control is no longer about micromanaging tasks. It is about owning the framework in which decisions are made. It means shaping the model's parameters, defining thresholds for intervention, and curating the feedback loop. In a hybrid environment, collaboration is not a surrender but a shared commitment that must be transparent and visible.

Research from the Harvard Business Review echoes this reframing. They argue that agentic AI works not by eliminating humans but by redefining how humans participate, moving from execution to evaluation, from doing to deciding.[8] This shift is not intuitive. It requires new leadership postures, where control is exercised through boundary design and oversight layers, not through direct manipulation.

The Hybrid Mind lives in this tension. It learns to trust automation while resisting complacency. It collaborates with machines not to offload responsibility but to amplify discernment. The paradox is not a flaw – it is a feature. Navigating it well is what separates augmentation from automation.

CHAPTER SUMMARY

Hybrid Intelligence is not a theory; it is an unfolding reality. Across industries and disciplines, human-AI collaboration is reshaping how decisions are made, how systems are designed, and what it means to lead. We are no longer in an era where answers alone suffice. The edge lies in context, orchestration, and the capacity to navigate ambiguity without surrendering to it.

This chapter traced the foundational shift: from humans as task executors to system orchestrators, from AI as a tool for answers to AI as a constructor of contexts, from monolithic models to modular reasoning, and from control as oversight to collaboration as design. Each shift invites a new kind of literacy – technical, emotional, and wholly human.

We have also acknowledged the cost of proximity. As systems become more capable, the impostor syndrome feeling grows more common. The Hybrid Mind does not ignore this. It accepts that augmentation, to be meaningful, must also be ethical. It makes space for doubt, slowness, and the kind of leadership that does not just manage systems but stewards meaning.

It is not a return to human supremacy or a surrender to machine autonomy. It is a convergence, a contract, a mindset. The Hybrid Mind does not fear being outpaced by machines. It stays close to what machines cannot hold – intuition, wisdom, and the burden of care. Furthermore, in doing so, it leads.

REFERENCES

1.   Joseph X. Ng and Andres Fortino, Strategic Innovation Governance: Identifying Crucial Use Cases via Experimentum Crucis Technology Matrix (ECTM) (New York: Decision Sciences Institute, 2025).
     Introduced to ground the chapter's core idea that orchestration is both a technical and ethical imperative in hybrid systems. The ECTM model supplies a governance framework for managing modular AI components, ensuring that automation aligns with strategic human oversight and enterprise values.
2.   Wharton School of the University of Pennsylvania, "Why Hybrid Intelligence Is the Future of Human-AI Collaboration," *Wharton Knowledge*, October 2024, https://knowledge.wharton.upenn.edu/article/why-hybrid-intelligence-is-the-future-of-human-ai-collaboration/.
     Quoted in Section 1.1 to define Hybrid Intelligence as "the computational strengths of AI fused with the holistic comprehension of human cognition." This concept supports the central argument that human-AI collaboration must transcend efficiency and reinforce trust and creativity.
3.   Joseph X. Ng, *EXAI Manifesto: Ethical eXcellence in Artificial Intelligence (EXAI) Manifesto* (New York: Humainority.org, 2025), https://Humainority.org/exai-manifesto/
     Cited throughout Chapter 1 to reinforce the principle of lifecycle accountability, which insists that human judgment must remain traceable across all stages of AI deployment.

The manifesto also provides ethical scaffolding for the design of context engines, modular inference, and hybrid collaboration boundaries.

4.   Data.World, "Context Engines Revolutionize AI," *Data.World Blog*, October 2024, https://data.world/blog/context-engines-revolutionize-ai/
      Used in Section 1.3 to define "context engines" as a key design pattern in AI architecture. These systems reconstruct the logical setting behind predictions, reinforcing the argument that AI should surface meaning and relevance, not just answers.

5.   Marvin Minsky, *The Society of Mind* (New York: Simon & Schuster, 1986).
      Introduced in Section 1.4 to establish the theoretical roots of multi-modular inference. Minsky's view of cognition as an emergent property of many simple "agents" supports the chapter's claim that modern AI should be constructed as interoperable modules – not monoliths.

6.   Shannon Vallor, *Technology and the Virtues: A Philosophical Guide to a Future Worth Wanting* (Oxford: Oxford University Press, 2016).
      Referenced in Sections 1.1, 1.5, and the conclusion to emphasize that the future of hybrid intelligence must be shaped by virtue ethics. Vallor's work lends moral grounding to the Hybrid Mind framework, particularly in how systems must support human flourishing, discernment, and moral imagination.

7.   Edward R. Tufte, *The Visual Display of Quantitative Information*, 2nd ed. (Cheshire, CT: Graphics Press, 2001).
      Referenced implicitly in Section 1.3 to justify the need for interpretability and ethical communication in AI output. Tufte's principles of data clarity and narrative integrity underpin the chapter's call for trustworthy insight presentation.

8.   David M. West and Rachel Goff, "What Is Agentic AI and How Will It Change Work?" *Harvard Business Review*, December 2024, https://hbr.org/2024/12/what-is-agentic-ai-and-how-will-it-change-work.
      Cited in Section 1.5 to illustrate how agentic AI systems shift human roles from execution to evaluation. The authors argue for intentional design pauses – or "ethical hesitation points" – which reinforce the chapter's argument that collaboration without agency is not augmentation but abdication.

9.   Luciano Floridi and Josh Cowls, "A Unified Framework of Five Principles for AI," *Nature Machine Intelligence* 1 (2019): 65–67.
      Referenced in the final paragraph to support the Hybrid Mind's commitment to values-driven decision-making. The five principles – beneficence, non-maleficence, autonomy, justice, and explicability – are aligned with the book's call for ethical, explainable, and accountable AI-human systems.

2 Understanding the OODA-F Loop

The OODA loop is introduced here as a foundational model for decision-making under pressure. Initially developed by John Boyd to give fighter pilots a cognitive edge in fast-moving combat, the loop – Observe, Orient, Decide, Act – has since been adopted across industries as a blueprint for agility.[1] However, in today's world, where intelligent systems now perceive, interpret, and act faster than humans, the classical OODA structure is under stress.

We propose its expansion into the OODA-F™ loop, with Feedback as a distinct and critical component that closes and calibrates the cycle (see Figure 2.1).[2] Across six sections, this chapter explores how each phase of the loop is reshaped by AI augmentation. Sensors, models, and data feeds have widened and accelerated observation, yet simultaneously distorted by overload and opacity.[3] Orientation – Boyd's most human phase – is now the most fragile, often displaced by predictive shortcuts and algorithmic framing.[4] Decision and action, once the domain of deliberative leadership, are increasingly compressed into automated sequences, creating the illusion of progress while bypassing ethical clarity.[5] Feedback emerges not merely as a reaction but as a layer of meaning and learning, essential for course correction and human trust.[6] The final section explores how we might restore the loop's rhythm by designing systems that preserve space for ethical hesitation and human tempo in a machine-driven world.[7]

The OODA loop is introduced as a theoretical upgrade and a practical and philosophical shift that enables hybrid human-AI systems to act quickly when it is safe and with intention when it matters (see Figure 2.2). This chapter lays the groundwork for the next: What happens when AI systems begin to run this loop independently?

2.1 ORIGINS OF THE OODA LOOP: BOYD'S FRAMEWORK OF AGILITY

The OODA loop – Observe, Orient, Decide, Act – was born in the cockpit. Developed by U.S. Air Force Colonel John Boyd, the framework originated to explain how

DOI: 10.1201/9781003648338-3

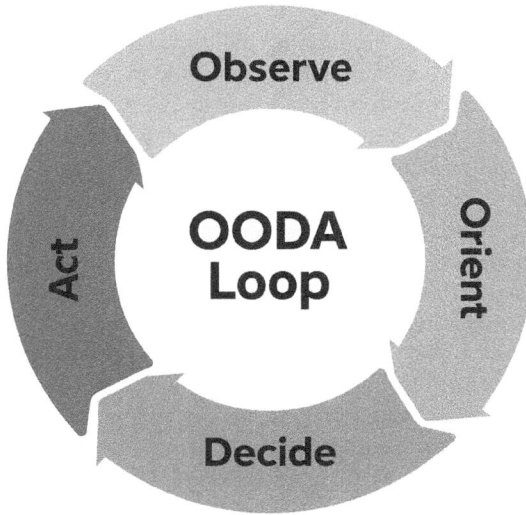

FIGURE 2.1   The OODA Loop. Image generated using DALL·E 3 via OpenAI (ChatGPT v4.0) based on author-provided design prompts. Reviewed and edited for accuracy.

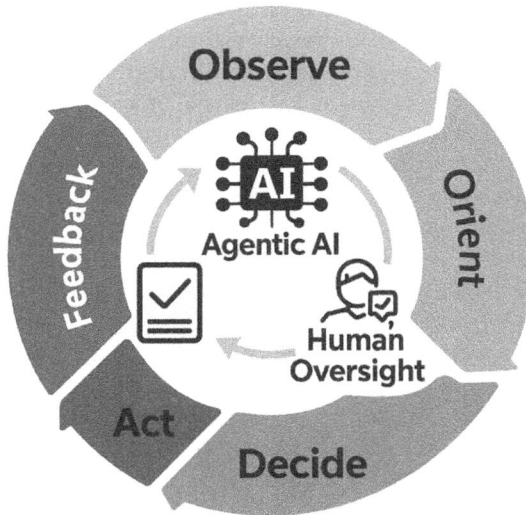

FIGURE 2.2   The OODA-F Loop – An Intelligent Feedback Ecosystem. Image generated using DALL·E 3 via OpenAI (ChatGPT v4.0) based on author-provided design prompts. Reviewed and edited for accuracy.

fighter pilots could gain a decisive advantage in air combat, not by brute strength or superior machinery, but by outthinking the opponent faster.[1] Boyd's insight was radical and straightforward: the individual who could cycle through decision-making loops faster and more adaptively would dominate the encounter, even if technically outmatched. It was not just a matter of military strategy but also an early recognition of temporal advantage as a competitive force. In modern terms, we might refer to it as a real-time strategy or rapid adaptation, but at its core, Boyd described cognitive agility in a dynamic environment.[8]

Boyd never published a formal book. Instead, he worked through extensive briefings, the most notable being Patterns of Conflict.[1] In that presentation, he reframed warfare as a contest of perception and speed, where disrupting the enemy's decision-making rhythm was more important than sheer firepower. While military applications were the starting point, it did not take long before strategists in business, cybersecurity, and systems engineering began borrowing the model.[9] Over time, the OODA loop became a shorthand for agility under pressure. What was once about dogfights in the Korean War is now a tool taught in MBA programs and product design sprints.[3]

It is worth noting that the loop is not linear. Boyd's recursive model loops back on itself, implying continuous adaptation rather than a one-time decision. Observation leads to orientation, which shapes decisions and drives actions, generating new observations in turn.[1] The speed and quality of this loop determine the effectiveness of the response. The loop has not become obsolete in today's environment – where information floods in through data streams, algorithms, and automated alerts. It has become overloaded. The problem is not having enough information to observe and discern what matters. The cycle has become asymmetrical. Humans still need time to orient themselves and make decisions, while machines quickly observe and act.[5] In this context, the OODA loop becomes the battleground of modern cognition.

We can see this tension in cyber defense. Systems that can "get inside" an adversary's loop, detecting intrusion attempts and responding before an attacker completes their own OODA cycle, have the upper hand.[3] In product design, agile methodologies echo the loop's rhythm: observe user behavior, orient toward insights, decide on iterations, and act through prototyping. Every primary tech firm cycle through these steps on a weekly, if not daily, basis. Boyd's influence is quietly embedded in the cadence of modern innovation.[3]

However, the weight of orientation is often missing in contemporary adaptations. While organizations rush to automate observation, decision-making, and action, the part Boyd considered most critical, orientation, is increasingly treated as an optimization problem.[8] This reduction misses the point. As we will explore next, orientation is where identity, intuition, experience, and emotional context live. It is the most human part of the loop, and the least replicable by AI.[4]

In recent applications, such as defense systems and autonomous vehicles, variations of the OODA loop have emerged that accommodate data streams and machine learning. These systems operate at machine speed but often at the cost of transparency.[5] What we gain in acceleration, we may lose in context. This chapter begins with the past because to evolve the loop, to hybridize it, we must understand what made it powerful in the first place: its recursive nature, its implicit invitation to adapt, and its human-centered design. It was never just a sequence. It was always a rhythm.[1]

2.2   OBSERVATION IN THE AGE OF DATA DELUGE

Observation used to mean looking closely. It required attentiveness, a practiced eye, and the discipline of noticing what others missed. In Boyd's original OODA loop, observation meant gathering real-world information, clues from the environment, movements from the opponent, and slight changes that signaled an advantage. However, observation is no longer a purely human act in the age of intelligent systems. AI-enhanced tools now watch everything, everywhere, all at once. From surveillance cameras and telemetry logs to user behavior analytics and biometric scanners, we have built systems that see more than we can process.[3]

It presents a paradox. Our interpretive clarity often shrinks as our observational capacities expand through machine augmentation. The sheer volume of incoming information, what many call the data deluge, threatens to flood the very decision cycles it was meant to empower. When every dashboard displays an urgent message, and every metric claims relevance, it becomes increasingly complex to discern what truly matters. AI, especially in its current generative and multimodal forms, often exacerbates this by producing summaries, predictions, or highlights that reflect correlations rather than comprehension. A model might point to a signal, but the burden of deciding its significance still falls on human orientation.[2]

The shift from active observation to passive data absorption introduces new fragilities. Humans risk losing their original perceptual instincts when they rely too heavily on machine-filtered views. In cybersecurity, for example, threat detection algorithms may highlight anomalies that look serious but are false positives. Over time, analysts may begin to trust the system's signals more than their skepticism – until the day something subtle slips through, simply because it was not statistically abnormal. Observation, in other words, is becoming increasingly indirect. We are watching interpretations, not reality.[6]

In creative work, the dilemma shows up differently. Consider the experience of a digital artist who utilizes generative tools like Midjourney or DALL-E to generate visual ideas. The tool offers dozens of options, often stunning and immediate. However, the more images it produces, the harder it becomes to trace the origin of one's vision. What did I see? What did I want before the machine told me what was possible? This is where Synthetic Syndrome begins to seep in, a quiet erosion of confidence not because we failed to observe but because observation was outsourced to the machine's eye.[10]

In business analytics, the same distortion appears when decision-makers are presented with algorithmically generated "insights" that are neatly packaged but contextually hollow. An executive may receive a weekly digest of sales trends, engagement rates, and predicted churn – impressive, fast, and often wrong. Because the system cannot observe intent, values, or nuance, the outputs reflect a pattern rather than the purpose. Here, the loop speeds up, but accuracy degrades. We have built faster observers, but not necessarily better ones.[3]

That brings us to the core issue for Hybrid Minds: AI augments our vision but refracts it. What we see now is always already filtered – by training data, by model weights, and by optimization goals set somewhere upstream. If we are not conscious of these filters, observation becomes an illusion. To regain clarity, we need what might be called meta-observation, the capacity to observe how we are observing. It involves situational awareness of the external world and the systems we use to perceive it.[2]

In battlefield environments, AI-enabled drones and real-time geospatial mapping have transformed observation from a linear input process into a multidimensional swarm of signals. However, even here, military strategists caution against over-automation. Too many alerts can be as dangerous as too few. This principle applies to corporate risk dashboards, automated compliance systems, and consumer-grade wearables. The signal is not the truth. It is a trace. The Hybrid Mind must learn to distinguish between the two.[5]

Ultimately, as we accelerate the Observe phase with intelligent systems, the question becomes: What are we giving up in return? Do faster observations lead to deeper understanding, or just more reaction? The power of the OODA-F loop does not reside in speed alone. It lies in the accuracy, meaning, and selectivity of perception. Without intentional filtering and human framing, observation is merely noise disguised as insight.[11]

2.3   ORIENTATION: THE HUMAN CORE OF CONTEXT

Orientation is not a technical function; it is a human act of sensemaking. While machines can observe and even mimic patterns of interpretation, orientation remains the most irreducible layer of the OODA-F loop.[1] This is where raw inputs meet lived experience – where instincts shaped by culture, memory, and emotional history determine what we pay attention to and how we assign value.[12] It is the invisible filter that turns data into direction.

Boyd called orientation the decisive phase, yet it is often overlooked in AI discourse, mistakenly regarded as mere data preprocessing or contextual tagging.[1] However, orientation is not statistical, it is personal. A threat to one group may be perceived as an opportunity by another. A pattern that appears clear through one cultural lens may dissolve when viewed through another.[4] This is where Human ETHOS lives, in the ambiguity, in the situatedness, in the refusal to collapse meaning into mere probability. Orientation is not just about forming a decision. It is about framing the very reality in which decisions are made.[7]

When generative AI offers options before you have finished a sentence, orientation becomes even more fragile. The system assumes the context; you inherit it.[10] This leads to what Synthetic Syndrome describes as interpretive erosion, a quiet displacement of our frameworks by algorithmic defaults. Over time, orientation becomes automated through templates, cues, and suggestions, until the human disappears not from the loop but from the meaning of the loop.[6]

Hybrid Minds resist this flattening. They slow the loop where it matters most, right here. Orientation is where leaders pause to ask: What story am I telling myself? What am I assuming? Who gets left out in this framing? In ethical dilemmas, orientation determines whether action is precautionary or extractive. In geopolitical crises, it defines whose pain is legible and whose is not.[13] In a job interview, it can mean the difference between bias and fairness. This is not optional metadata. This is moral infrastructure.[4]

AI can help us surface alternative interpretations, what might be called "modular viewpoints," but it cannot replace the process of choosing one over another.[14] That choice requires identity, empathy, risk tolerance, and historical awareness.[4] These are not bugs in human reasoning. They are the very signals that orientation encodes. In an age of acceleration, the Hybrid Mind returns to orientation not as a bottleneck but as the source of ethical calibration. Machines can compute, but only humans can contextualize.[9]

Orientation, then, is the axis of autonomy. It is where reflection defies reflex. It is not a delay in the loop; it is the loop's conscience.[7]

2.4   DECISION AND ACTION: THE ACCELERATION TRAP

In the classic OODA loop, once a person observes and orients, the next logical move is to decide and act. This sounds simple, see the problem, understand the context, make a choice, and act. However, in high-stakes environments, and especially in AI-augmented systems, the Decide-Act phase is increasingly distorted by what we now refer to as the acceleration trap.[3] It is the assumption that faster decisions are always better decisions. That speed equals superiority. However, without a grounding in understanding, speed is merely momentum without direction.

Traditionally, decision-making involves deliberation. Leaders weighed trade-offs, imagined second-order consequences, and reflected on alignment with long-term goals. The action came with responsibility. However, in AI-powered settings, that calculus has shifted. Tools now generate decisions, or at least decision suggestions, automatically. Recommender systems, predictive analytics, and optimization engines all nudge the user toward a pre-modeled outcome. In finance, an algorithm might flag a security for sale based on volatility thresholds. In HR, it may automatically rank résumé based on keyword fit. In medicine, it may propose treatment paths based on similarities between patients.[8] The decision feels fast, sometimes instantaneous. However, what is often missing is the reason it was created.

This is where the trap reveals itself. AI does not just accelerate the act of deciding, it redefines the shape of the decision.[5] It simplifies, rounds off its edges, and makes it more digestible. Furthermore, it quietly marginalizes the human process of wrestling with ambiguity. Over time, humans may stop deciding altogether. We approve rather than choose. Action becomes a formality, not a judgment.

This is not hypothetical. In multiple case studies, organizations that implemented AI decision-support systems saw a decline in the engagement of domain experts.[15] For instance, in logistics operations, AI route planners reduced the workload for dispatchers, but also eroded their situational awareness. When weather patterns, fuel costs, or regional disruptions shifted, many teams lacked the instinct to override the system. They acted too quickly because they had outsourced the responsibility to a process, they no longer fully understood.

In military applications, this trap has even more severe implications. Autonomous drone systems can identify and lock onto targets with minimal human input.[5] The technical decision might be sound, but what about the ethical one? What happens when a system can act faster than a human can morally process? If orientation is rushed or superficial, then the decision becomes a function of computational momentum rather than intentional judgment. In this case, action risks becoming a mere reaction.

This raises a core challenge for the Hybrid Mind: when to defer to speed and when to intervene. There are moments when rapid action is necessary, such as during disaster response, cybersecurity incidents, or trading anomalies. However, even in these cases, the decision must emerge from a structure that has taken human values into account from the outset.[2] In hybrid systems, one emerging solution is the use of intention gating, a design pattern where a human must explicitly authorize a particular class of high-impact decisions before they are executed. This restores friction not as inefficiency but as moral design.

Still, we must reckon with the psychological impact of speed. The more decisions are automated, the more human actors begin to question their judgment.[6] Why deliberate when the machine is already right? This is where Synthetic Syndrome reappears in a subtler form: the diminishing of self-trust, the erosion of decisional confidence in people who once prided themselves on sound judgment. Over time, even leadership may begin to favor compliance over conviction, because to decide is to risk being wrong, and AI systems rarely doubt themselves aloud.

The Hybrid Mind accepts acceleration, it tempers it. It acknowledges that in an era of automated action, choosing when not to act is just as important as executing quickly. Decision and action remain core parts of the loop, but their quality depends entirely on what comes before, and what comes after.

2.5   FEEDBACK: FROM OUTCOME TO CALIBRATION

Feedback is the heartbeat of any adaptive system. It tells you not only what just happened but whether what happened should happen again. In Boyd's original formulation of the OODA loop, Feedback was implied rather than isolated – it flowed back from Act to Observation in a continuous loop.[1] Nevertheless, in today's world of intelligent systems, automation, and human-AI collaboration, Feedback deserves explicit status as its stage. Not just a byproduct of action but a design priority.[3]

At its most basic, Feedback is information about the consequences of a decision. However, Feedback also serves as a means of correction, reinforcement, and reflection. It helps us recalibrate our orientation, update our mental models, and learn from our errors.[9] In AI systems, Feedback is built into the architecture. Machine learning models are constantly adjusted through loss functions, gradient descent, and human labeling. Reinforcement learning agents evolve by receiving signals from the environment that reward or penalize their actions. These technical mechanisms echo how humans learn, but they lack emotional weight, moral nuance, and situational complexity.[8]

In human systems, Feedback is more layered. It is not just telemetry or metrics. It is emotion, perception, reputation, and ethics. For instance, a manager might implement a new workflow based on AI recommendations and receive Feedback in the form of plummeting team morale, missed nuances in edge cases, or customer frustration. These are signals, too, but they are not easily captured in dashboards. They require interpretation, empathy, and judgment.[2] That is why, in hybrid systems, we need a broader definition of Feedback: not just what the system captures, but what the human senses.[7]

There are three common failures in feedback loops. The first is latency: Feedback comes too late to matter. A quarterly review will not help with a decision that was made and forgotten months ago. The second is opacity: the Feedback is available but obscured behind metrics that do not match the original intent. A drop in churn might appear to be a success, but it could be masking more profound dissatisfaction. The third is asymmetry: some stakeholders experience the consequences of decisions but have no channel to feed that experience back into the system. This is especially common in healthcare, education, and HR, where system outcomes significantly impact people, but individuals have limited influence over system redesign.[5]

To address these issues, hybrid systems must incorporate an explicit feedback architecture. That means building channels, technical, cultural, and emotional, that capture not just results but meaning. In advanced implementations, this looks like reflexive AI systems that integrate Feedback across user behavior, outcome audits, and sentiment analysis.[15] In simpler terms, it looks like organizations are asking: "Did this work? For whom? At what cost?" These questions must be institutionalized, not improvised.[11]

For example, Google's AI-based cybersecurity tools now incorporate feedback loops not just through event logs but through red team adversarial testing, analyst commentary, and incident response debriefs.[14] This multilevel Feedback helps reorient the system not just toward accuracy but toward resilience. In the world of hiring and job-matching platforms, bias feedback, captured from user experience, legal review, and performance disparities, now informs real-time algorithmic adjustments. These are signs that Feedback is evolving from passive monitoring to active recalibration.[10]

There is also a personal side to Feedback. Many knowledge workers report a loss of internal barometers when their roles become AI-augmented. They do not always know if they made the right choice; they only know if the AI was followed. This is another facet of Synthetic Syndrome, when the erosion of Feedback reduces self-trust.[6] Over time, people stop asking, "How did I do?" and start asking, "Did the system approve?" This inversion is quiet but consequential.

For the Hybrid Mind, Feedback is not just a loop, it is a mirror. It reflects our assumptions, exposes blind spots, and asks us to change.[12] It is both technical and ethical. It is how we stay accountable, not only to performance metrics but also to our purpose. As we build faster systems, we must also build slower reflections. Systems that pause, ask, listen. And change.[16]

2.6   RESTORING THE LOOP: HUMAN-AI RHYTHM

If the OODA-F loop has become faster, more complex, and more machine-driven, then restoring it is not about slowing everything down, it is about rebalancing the rhythm.[12] In hybrid systems, the challenge is not technological sophistication; rather, it is the integration of multiple systems. It is tempo. The loop is no longer purely cognitive; it has become procedural, algorithmic, and institutionalized.[9] The rhythm once governed by human awareness has been hijacked by systems that rarely hesitate to act. However, it is in the hesitation that ethics live.[4]

To restore the loop, we must recognize that human-AI collaboration demands more than efficiency. It demands pacing. Observation, orientation, decision, action, and Feedback must unfold in a tempo that allows for alignment – not just with organizational goals, but with human dignity and contextual understanding. Where the loop runs too fast, meaning is lost. Where it runs too slow, opportunity fades. The Hybrid Mind is not simply fast – it is tempo-aware. It adjusts the pace based on stakes, ambiguity, and consequences.[6]

In military contexts, this principle is already being reexamined. Decision-support systems that operate on drone telemetry or satellite imagery can process, recommend, and act on targeting data in seconds.[5] However, ethical oversight does not operate on the same clock. Moral clarity takes time. The legal review takes deliberation. This is why many armed forces now require human-in-the-loop confirmation for lethal actions.[5] These are not just safeguards – they are interruptions designed by nature. A strategic pause is not a failure; it is a deliberate choice. It is a feature.[5]

We see similar dynamics in healthcare. AI diagnostic tools, such as IBM Watson or Google's DeepMind-powered systems, can scan radiology images faster than any human clinician.[2] However, the act of diagnosis is not only about detection – it is about trust, context, and communication. A system that can identify cancer but cannot explain its reasoning cannot operate alone. The human interruption must be preserved, not as a fallback, but as a layer of responsibility. The loop needs space to breathe.[2]

This is why restoration of the loop begins with design. Systems must be built with ethical hesitation mechanisms – moments where human agency is reinserted.[4] In product development, this might take the form of escalation thresholds. In AI safety, it may look like model interpretability checkpoints. In governance, it becomes layered accountability. However, in every case, the goal is the same – to give humans enough time to matter.[7]

We might also think of this rhythm as a form of "decision latency tuning." In AI-driven markets, where microseconds separate winners from losers, latency is the enemy. Nevertheless, in governance, strategy, and justice, latency is the space for values to enter. The Hybrid Mind must learn to toggle between minimizing latency and preserving latency. Not everything should be frictionless. Some things should be hard. Some things should take time.[9]

One-way organizations are achieving this is through layered loops. Instead of running one flat OODA cycle, they create nested rhythms: fast loops for automation, mid-speed loops for managerial oversight, and slow loops for ethical review and analysis of long-term consequences.[11] These layers create a buffer, a rhythm of rhythms, where human insight can interrupt the algorithmic flow when it truly matters.[14]

Nevertheless, restoration is not just a procedural matter, it is a cultural one. People in the system must feel authorized to intervene.[10] The most advanced feedback architecture in the world is useless if no one feels safe enough to challenge the output. Psychological safety becomes a design prerequisite. This means training for interruption, valuing reflection, and rewarding doubt. The Hybrid Mind does not just interpret. It intervenes.[6]

At its core, this chapter is a reminder: automation is not destiny. Rhythms can be reset. Loops can be humanized. Furthermore, in doing so, we accept machine intelligence, we shape it.[16]

CHAPTER SUMMARY

The OODA-F loop – Observe, Orient, Decide, Act, Feedback – is not just a decision model. It is a lens for understanding how humans and intelligent systems can navigate uncertainty together.[1] What began as a military tool for achieving aerial dominance has evolved into a universal framework for adaptive strategy.[12] Moreover, in the age of AI, it is evolving again, not by growing more complex, but by demanding more profound clarity at each phase.[3]

As we have seen, observation is no longer constrained by human perception. Machines extend our senses but also shape what we see and what we overlook.[2] Orientation remains deeply human – a fusion of intuition, experience, and cultural memory – but algorithmic shortcuts and pattern substitutions increasingly threaten it.[7] Decision and action are now wrapped in automation, where speed tempts us to confuse output with insight.[5] Feedback, once an afterthought, must now be elevated to a structural component, a deliberate loop of learning, correction, and meaning restoration.[6]

What binds these stages together is not just data or logic. It is rhythm.[12] The Hybrid Mind thrives not by outpacing machines but by sensing the right tempo for judgment, pause, and presence.[6] Restoring the loop is not about slowing down, it is about knowing when to. It is about preserving space for ethical interruption, emotional resonance, and human accountability.[4]

The expanded OODA-F loop offers a compass in systems where the lines between automation and agency are blurred.[9] It gives leaders and designers a structure for asking the right questions at the right time. It reminds us that strategy is not just about speed but also about coherence.[14] Furthermore, it signals that in the convergence of man and machine, the most powerful moves may be the ones that require us to stop, reflect, and reframe.[16]

To lead in a world of intelligent systems is to master this loop, not as a formula but as a practice. The Hybrid Mind does not just move through the loop. It listens to it.[15]

REFERENCES

1. John Boyd, *A Discourse on Winning and Losing* (unpublished briefing slides, 1986). Boyd's OODA loop emerged from aerial combat strategy, emphasizing mental agility and recursive adaptation. It laid the groundwork for viewing decision-making as a dynamic and competitive cognitive process.

2. Joseph X. Ng and Andres Fortino, "Strategic Innovation Governance: Identifying Crucial Use Cases via Experimentum Crucis Technology Matrix (ECTM)" (New York University, 2025).
This paper provides the structural rationale for evolving the OODA loop into the OODA-F loop, formalizing Feedback as a distinct stage in hybrid decision architectures.
3. Michael Chui, James Manyika, and Mehdi Miremadi, *The Executive Guide to AI* (McKinsey & Company, 2018).
This guide helps contextualize how AI reshapes observation and decision-making processes, while also identifying the risks of over-compression and blind automation.
4. Shannon Vallor, Technology and the Virtues: A Philosophical Guide to a Future Worth Wanting (Oxford University Press, 2016).
Vallor's virtue ethics framework underpins the chapter's call for restoring human pacing, moral reflection, and design features that support ethical hesitation in high-speed systems.
5. Jacques Bughin et al., "Notes from the AI Frontier: Modeling the Impact of AI on the World Economy," *McKinsey Global Institute* (2019).
This report outlines how automation distorts orientation and accelerates execution without ensuring interpretability, reinforcing the chapter's concern that speed can undermine clarity.
6. Cathy O'Neil, Weapons of Math Destruction: How Big Data Increases Inequality and Threatens Democracy (New York: Crown Publishing, 2016).
O'Neil's work supports the chapter's claims about feedback asymmetry, decision opacity, and the potential for systemic harm when feedback loops are misaligned with human experience.
7. Eric Horvitz, quoted in *The Hybrid Mind: The Human-AI Convergence*, 2025.
Horvitz advocates for mixed-initiative systems, offering insight into why ethical hesitation and interpretability must be preserved as AI accelerates decision cycles.
8. C. Dumitru, "Crucial Instances and Crucial Experiments in Bacon, Boyle, and Hooke," *Society and Politics* 7, no. 1 (2013): 45–53.
This historical analysis supports the chapter's framing of the OODA loop as a model of epistemic acceleration and selection, akin to the philosophical logic of crucial experiments.
9. Robert Coram, *Boyd: The Fighter Pilot Who Changed the Art of War* (New York: Little, Brown, 2002).
Coram expands on Boyd's influence, showing how the OODA loop transitioned from military doctrine into broader strategic fields including business and leadership.
10. Sherry Turkle, Reclaiming Conversation: The Power of Talk in a Digital Age (New York: Penguin Press, 2015).
Turkle's analysis of reflective silence and emotional presence in digital systems contributes to the chapter's call for tempo-aware systems that include space for human judgment and challenge.
11. Yuval Noah Harari, *Homo Deus: A Brief History of Tomorrow* (New York: Harper, 2017).
Harari warns of society's growing trust in algorithmic authority and the erosion of interpretive autonomy – ideas that align directly with the chapter's critique of synthetic decision-making rhythms.
12. Howard Gardner, *Five Minds for the Future* (Boston: Harvard Business School Press, 2007).

Gardner's "ethical mind" and "respectful mind" concepts reinforce the moral dimensions of orientation and feedback in the OODA-F framework, especially as they relate to cultivating long-term decision wisdom.

13. Luciano Floridi and Josh Cowls, "A Unified Framework of Five Principles for AI in Society," *Nature Machine Intelligence* 1, no. 1 (2019): 65–67.
Floridi and Cowls offer a principled model for embedding responsibility and justice into AI systems, aligning with the chapter's emphasis on feedback as an ethical mechanism.

14. Peter Voss, *Cognitive AI Research Papers and Notes*, Adaptive AI Inc., 2005–2023.
Voss's modular cognition theory supports the architectural layering of loops and feedback stages, reinforcing the design logic behind the OODA-F loop as an adaptable, interpretable framework.

15. David L. Sackett et al., *Evidence-Based Medicine* (Churchill Livingstone, 2000).
Sackett's focus on clinical reasoning reinforces the necessity of calibrated decision timing and empirical feedback, validating the chapter's argument for intentional pauses in hybrid systems.

16. Deloitte, State of AI in the Enterprise: 5th Edition (Deloitte Insights, 2023).
This industry report offers empirical backing for the chapter's critique of real-world AI systems that fail due to insufficient feedback infrastructure and the absence of human-in-the-loop review mechanisms.

3 Rethinking the OODA Loop

Human Decision-Making in an AI World

Marking a pivotal shift in The Hybrid Mind, this section transitions from exploring human-AI collaboration to examining the systems that now govern decision-making. At its core is the transformation of John Boyd's OODA loop in the context of AI acceleration.[1] Initially a military strategy for adaptive superiority, the loop has now evolved into the foundation of modern intelligent systems. However, as algorithms compress time and transform orientation into data patterns, the original loop is no longer sufficient. This chapter introduces a crucial evolution: OODA-F. Here, feedback is not just a trailing consequence but a formal, visible phase, a key ethical design feature that restores human judgment in the loop.[2]

This chapter examines how intelligent systems, particularly agentic AI, operate at a pace that exceeds their creators' comprehension.[3] These systems observe, orient, decide, and act independently, but without a structured feedback phase, they risk acting without reflection. From healthcare triage to cybersecurity, the stakes are high: compressed loops without human override mechanisms invite ethical blindness, systemic fragility, and adversarial exploitation.[4] The Hybrid Mind must resist this acceleration, not through withdrawal, but through rhythm.[5] Ethical hesitation,[3] loop defense,[6] and participatory recalibration[2] become crucial human contributions to system architecture.

What emerges is not a critique of speed but a redefinition of responsibility.[7] The Hybrid Mind's role is not to override automation out of fear but to compose its tempo.[8] In this new orchestration, the loop does not close with action, it closes with meaning.[9] That is the work of OODA-F. It sets the stage for Chapter 4, where we examine Agentic AI at its core: the design of systems that not only make decisions but also learn to account for their own decisions.[10]

3.1 AGENTIC AI AND THE ETHICS OF INTERVENTION

In a world increasingly mediated by algorithmic acceleration, the foundational military framework known as the OODA loop – Observe, Orient, Decide, Act – has begun to fracture under the pressure of automation. What John Boyd initially conceived as a dynamic decision-making cycle for air combat now struggles to hold its shape in systems where actions are taken not just faster than human reaction times but

DOI: 10.1201/9781003648338-4

sometimes without any human presence at all.[1] It is not merely a technical issue. It presents both philosophical and governance challenges. The growing presence of Agentic AI, systems designed with semi-autonomous feedback loops, has transformed the OODA loop into something else entirely. Something faster. Something riskier. Something that needs a fifth phase: Feedback.

Agentic AI systems differ fundamentally from traditional automated tools. They do not just respond, they infer, adapt, and choose.[11] These systems are embedded with internal goals, prioritization heuristics, and recursive calibration mechanisms. In cybersecurity, fraud detection, autonomous vehicles, and battlefield simulations, agentic systems monitor environments in real time and update their strategies without awaiting human command. While this confers undeniable advantages in speed and scope, it also introduces new ambiguity: Who decides when the system has gone too far or too fast? As outlined in the EXAI Manifesto, feedback is not a soft afterthought, it is an architectural necessity, particularly when systems increasingly act before we can observe them.[2]

This tension is most evident in the erosion of the "Orient" phase, the part of the loop intended to integrate human judgment, lived context, and moral reasoning. AI accelerates time and condenses decisions, but what it often overlooks is the moment of reorientation: the pause, the perspective, the hesitation. Orientation is not just data fusion – it is meaning-making.[6] When AI systems collapse this phase into decision and action, the human is reduced to a supervisor at best or an after-the-fact explainer at worst. What we gain in speed, we may lose in coherence.

That coherence is especially fragile when systems operate in adversarial environments. A cybersecurity agent trained to neutralize emerging threats might act on predictive signals without complete transparency. A trading algorithm could respond to volatility by withdrawing extreme liquidity, thereby compounding market instability. These are not science fiction scenarios; they are documented concerns. As the Strategic Innovation Governance paper explains, autonomy without interpretability becomes fragile, especially when innovation outpaces institutional safeguards.[4]

The solution, however, is not to abandon autonomy, but to structure it around agentic feedback. Feedback in OODA-F is not the end of the cycle but a looping anchor, a formal recognition that any intelligent system, no matter how fast or optimized, must build in a channel for recalibration, counterfactuals, and dissent. It is in feedback that orientation is reclaimed. Moreover, it is here that the Hybrid Mind must operate not to outcompete the machine but to recenter judgment when judgment has gone missing.

Consider the challenge of AI in healthcare triage. A model might flag a patient as low risk for readmission based on thousands of data points but overlook cultural, linguistic, or psychosocial factors that are not well represented in the data. Without structured feedback from the clinician, who understands the patient's narrative, the model perpetuates blind spots under the guise of objectivity. Feedback is not simply a correction mechanism; it is a safeguard against erasure. As seen in Synthetic Syndrome, the erosion of human confidence in such contexts arises not because people doubt their competence but because their contextual insight no longer feels relevant in a world of instant answers.[6]

This is why OODA-F is not a mere academic update. It is a design intervention. In the Multi-Modular Inference architecture, feedback loops operate not only between humans and machines but also across AI modules, between SLMs that surface micro-contexts and LLMs that generate narratives. Each loop must close not only with action but with re-evaluation. Who should speak next? Was this the proper context? Did the user's intent shift mid-stream? These are not afterthoughts; they are moments where agency is either preserved or lost.[11]

Agentic AI must be constrained not by distrust but by design.[9] The moral brake, the moment of ethical hesitation, is no longer a luxury. It is the most human part of the system. Feedback serves as a reminder that speed is not wisdom and that orchestration without reflection is merely noise. The OODA-F loop, then, is the new battleground for hybrid decision-making: one where intervention is not just possible but necessary. One where the Hybrid Mind does its real work, not by matching the machine in velocity, but by restoring the rhythm of reason.

3.2   HUMAN OVERRIDE AND ETHICAL HESITATION

As intelligent systems evolve from assistants to agents, the need for human override becomes not just a technical safeguard but a moral obligation. In AI-driven decision cycles, especially those governed by speed and scale, the question is no longer if systems should act independently, but when and under what conditions they must be interrupted. Ethical hesitation, as a practice, emerges from this tension. It is the intentional slowing of decision momentum, not because we distrust intelligence but because we recognize the limits of its judgment.[3]

This is not a hypothetical concern. In defense sectors, autonomous systems are already making life-and-death decisions in simulation environments. Department of Defense Directive 3000.09 mandates that lethal autonomous weapons must include meaningful human control, yet this phrase, "meaningful," remains underdefined in both technical and philosophical terms.[12] How long must a human have to intervene? What degree of system transparency is necessary to exercise that override meaningfully? As explained in the EXAI Manifesto, systems must be designed to invite hesitation, surfacing the confidence in their inferences, the provenance of their data, and the degree of uncertainty involved in their conclusions.[2]

The emotional terrain here is just as important as the logical one. Synthetic Syndrome, the modern erosion of human confidence in AI-augmented workplaces, reveals that hesitation is becoming culturally stigmatized.[6] In industries driven by metrics and optimization, pausing to ask a question may feel like a failure. However, in a world where AI can produce confident, coherent-sounding outputs without understanding, that pause may be the only ethical act left. Ethical hesitation is not indecision; it is resistance to premature closure.[9]

Consider an AI system trained to triage patients in the emergency room. It processes vital signs, medical history, and symptom data faster than any physician. However, it may overlook domestic abuse indicators, language barriers, or subtle emotional cues. A clinician's override, based not on the AI's data but on factors such as eye contact or silence, maybe the only thing that saves a life. This override is not a correction; it is an addition to the system's logic. It injects context, care, and culture into the loop.[4]

The Strategic Innovation Governance framework emphasizes this human checkpoint as a design priority, not a contingency. Systems must be designed to reveal when and where their decisions cross thresholds of ethical ambiguity, what EXAI refers to as the "red zones" of augmentation. These are areas where no amount of data suffices and where only lived experience or communal values can arbitrate.[13] Without clear channels for override, intelligent systems become brittle, blind to the nuance they were never built to see.

In agentic architectures, where AI modules act independently and asynchronously, overrides must also be distributed. The Multi-Modular Inference model shows that different layers of AI logic (e.g., symbolic, statistical, contextual) may conflict. If one model detects fraud and another deems it safe, who or what arbitrates the final decision? Human override must extend beyond interface-level control and into system-layer mediation. This is the essence of orchestration: not suppressing AI action but composing its possibilities within the boundaries of human values.[14]

Importantly, override mechanisms are not only for edge cases. They represent a cultural assertion: that human experience, intuition, and even doubt are not inefficiencies but features of ethical systems.[8] Slowness, in this context, is a design affordance. It ensures that in moments of gravity, humans can insert not just logic but empathy.

Ethical hesitation, then, is not just a pause, it is a posture. It is a declaration that just because a system can act does not mean it should. Furthermore, just because a machine is confident, it does not necessarily mean it is right. In Hybrid Mind design, hesitation is where humanity makes its presence felt.

3.3 ADVERSARIAL LOOPS AND LOOP DEFENSE

All systems learn. However, not all systems learn safely. In adversarial environments, where actors deliberately manipulate inputs to induce system failure, the traditional OODA loop encounters a unique vulnerability: it can be hijacked. Intelligent systems, especially those operating in semi-autonomous or agentic configurations, are susceptible to feedback poisoning, input deception, and model inversion. These threats do not just attack the AI's decision quality, they attack the loop itself. In doing so, they threaten to replace orientation with distortion, decision with reaction, and action with misfire.[6]

Adversarial AI is not limited to sophisticated cyberattacks. It can manifest as minor image perturbations that fool computer vision models, algorithmic manipulation of recommender systems, or misinformation loops in social media engines. In each case, the system's learning becomes a weapon against itself. For instance, a content recommendation engine trained on click-through rates may amplify divisive content, not out of malice but because it has been optimized to interpret outrage as a form of engagement. Over time, the loop is no longer observing the world, it is observing its distorted output. Orientation collapses into echo.[15]

The problem deepens in agentic architectures where the loop is recursive. The system acts, observes the consequences of its actions, and learns from those consequences. In adversarial contexts, the feedback is no longer neutral. It is shaped, sometimes subtly and sometimes overtly, to induce specific behaviors. In cybersecurity, attackers have utilized this approach to significant effect, by gradually introducing benign-looking anomalies to fraud detection systems until the model learns to normalize malicious behavior.[11] This is not just an attack on data. It is an attack on trust.

Loop defense begins by acknowledging this vulnerability. It requires treating the Feedback phase of the OODA-F loop not as a passive receiver of outcomes but as a target in need of verification. Feedback, when unchecked, becomes a liability. Nevertheless, when made transparent, traceable, and adversarially aware, it becomes a line of defense. In the Strategic Innovation Governance framework, this is described as loop hygiene. This governance layer regularly evaluates whether the system's feedback mechanisms are being shaped by noise, manipulation, or market distortion.[13]

Consider the case of financial trading models that learn from other AI-driven competitors. These systems often enter a reflexive loop, where each system responds to the anticipated response of another. Like a hall of mirrors, the signal is lost in the anticipation. What appears to be emergent intelligence is sometimes just recursive confusion. In extreme cases, this dynamic has caused flash crashes, where liquidity disappears in seconds because systems "learn" that everyone else is about to exit.[16] These are not bugs. They are the logical consequence of loops without human governors.

In this context, the Hybrid Mind must play the role of loop auditor. It does not merely act or observe, it tests the integrity of the system's logic. Are the signals being gamed? Are the inferences being shaped by outside influence? Are the outcomes reinforcing the wrong behavior? Loop defense is not reactive patchwork; it is an ongoing design posture. One that requires humans to intervene not only when outcomes fail but also when assumptions drift.

That drift, subtle, cumulative, and often unnoticed, is what makes adversarial loops so dangerous. It is not that the system breaks; rather, it gradually ceases to serve its intended purpose. In AI ethics, this is referred to as value misalignment. Nevertheless, in hybrid leadership, we must recognize it as loop erosion, the degradation of signal clarity over time. The antidote is not to distrust all feedback but to challenge it, to embed friction where necessary, skepticism where useful, and recalibration as routine.[9]

The role of the Hybrid Mind is not to beat adversaries at their own game but to protect the integrity of the system's feedback structures.[17] Because when a system no longer knows what it is learning from, it no longer knows what it is becoming.

3.4   DESIGNING ETHICAL FEEDBACK: THE FIFTH O IN OODA-F

Feedback has always been present in the OODA loop – an implied afterthought tucked into the transitions between Act and Observe. However, in the era of autonomous systems and recursive learning, feedback cannot remain implicit. It must be elevated – designed, structured, and made visible. In the OODA-F model, feedback is not just the fifth phase; it is also the fourth phase. It is the ethical anchor – the moment where acceleration pauses, not in weakness, but in reflection. Feedback transforms the loop from a closed circuit into a civic process.[2]

Most AI systems today operate with minimal feedback governance. They register accuracy, optimize for efficiency, and self-correct within the boundaries of their training data. However, they do not ask: Was this outcome just? Did this decision preserve dignity? Was harm reduced? These questions are not part of traditional loss functions. They require a different kind of loop, one that brings human values into the circuit. As Fei-Fei Li has noted, AI must be designed to enhance human capability, not just replicate performance.[5] Moreover, that enhancement is only possible when systems are held accountable to feedback beyond the metric.

Designing feedback, then, begins with intent. What kind of information should the system seek after acting? Who gets to provide that feedback? How is it used, retained for transparency, integrated into learning, or flagged for human review? The EXAI Manifesto frames this as the right to participatory recalibration – a design principle that enables affected individuals or communities to identify and address invisible impacts and correct systemic drift.[2] In this way, feedback becomes democratized. It ceases to be a technical variable and becomes a civic instrument.

Take the example of a hiring algorithm trained on historical success patterns. The system may "learn" that certain schools, names, or employment gaps are associated with poor outcomes. Without intentional feedback from applicants, recruiters, or third-party auditors, these biases perpetuate themselves in silence. Feedback mechanisms, if present, are usually reduced to binary performance flags or quiet opt-outs. This is not feedback; it is frictionless injustice. A proper OODA-F system would not only register who was rejected but also ask why, what context was missed, and who was affected. It would learn not just from performance data but also from disagreements.[6]

This type of feedback requires what the Strategic Innovation Governance paper refers to as adaptive orientation scaffolds, systems designed to adjust orientation pathways when feedback reveals that the original assumptions no longer apply.[4] In practice, this might involve requiring models to disclose their confidence intervals, trace their input sources, or ask users whether the outcome aligns with their lived experience. Such scaffolds are not just guardrails. They are re-education mechanisms for the system, and sometimes for the human users as well.

Importantly, not all feedback is created equal. In adversarial or high-volume environments, signal contamination becomes a risk. This is why feedback must be auditable, layered, and contextual. A loop that learns from unverified, manipulated, or malicious signals will inevitably erode. As Eric Horvitz of Microsoft observed, "Humans and machines are not adversaries, but teammates – each with different strengths, combining for greater outcomes."[7] Nevertheless, for that collaboration to be real, machines must learn to listen, and not just to data.

In the Hybrid Mind model, feedback is where co-authorship happens. It is the phase where the human mind does what machines still struggle to do: reflect across domains, weigh meaning over speed, and synthesize values into pathways forward. Feedback is also where Synthetic Syndrome softens, where humans see that their judgment still matters, not because it is faster or louder, but because it is ethical.[6]

OODA-F is not just a redesign of a strategic loop. It is a reassertion of humanity. Feedback is how we remember that intelligent systems are not only tools of inference but participants in a social contract. Moreover, the Hybrid Mind is its witness, editor, and – when needed – its interruption.

3.5 RHYTHM, RESISTANCE, AND THE HYBRID MIND'S ROLE IN THE LOOP

The modern decision environment is not slow. It pulses. Systems cycle through decisions in microseconds. Algorithms retrain. Agents recalibrate. Newsfeeds refresh. In this landscape, the Hybrid Mind is not the fastest entity. It is the most intentional. Moreover, that intention manifests not in dominance over machines but in rhythm – knowing when to slow, when to listen, and when to intervene. In the OODA-F loop, this rhythm becomes the Human signature.[2]

Rhythm, in this sense, is not aesthetic – it is operational. It is the difference between responding and reacting. Fast systems tend to default to a reactive approach, especially under pressure. However, reaction is not always the highest form of intelligence. The Hybrid Mind resists the compulsion to match AI's velocity. It embraces asymmetry. Where the system loops rapidly, the human punctuates. The Hybrid Mind's power lies in knowing that intelligence is not a sprint, but a syncopation.[8]

That syncopation shows up in pauses. In the moment a leader reviews a model's output and asks, what is missing? In the way a team reruns a scenario, not for performance, but for ethical clarity, in how a community insists that systems be interpretable – not just functional. Resistance in this context is not obstruction. It is design justice. It is the refusal to let speed become an excuse for oversight failure. It is the insistence that context matters more than convenience.[9]

Hybrid thinkers are not just users of intelligent systems – they are stewards of the loop. Their job is not to override constantly but to reintroduce cadence into feedback-deprived architectures. When a machine completes the OODA-F loop 100 times before a human has time to blink, it is the Hybrid Mind that imposes a meta-loop: a rhythm of review, recalibration, and restraint. This is not just operational hygiene – it is moral architecture.[11]

What this demands is not more data or faster tools but better reflexes. Reflexes that are not defined by urgency but by empathy. The Hybrid Mind must hold competing tensions: the drive to move forward and the wisdom to wait. The desire to optimize and the imperative to humanize. These tensions do not cancel each other; they shape the loop. They give it its distinctive rhythm, one that beats to the tempo of trust.[5]

Moreover, in that rhythm lies resistance. Not the resistance of defiance but of discernment. The Hybrid Mind must resist the erasure of meaning through automation. Resist the flattening of nuance through model compression. Resist the temptation to act simply because the system says "now." Instead, it must insert moments of why.[6]

As Synthetic Syndrome has shown, this kind of pause can feel powerless, especially in a world where machines generate output faster than humans generate questions. However, that pause is sacred. It is the moment when the loop becomes legible, when power is shared, and where leadership – real, accountable, and hybrid – shows up.[6]

OODA-F is not just a framework. It is a rhythm map. Furthermore, the Hybrid Mind is its conductor.

CHAPTER SUMMARY

The OODA-F loop – Observe, Orient, Decide, Act, and Feedback – emerges not as an abstract framework but as a living system architecture redefined by AI acceleration.[2] In this chapter, we reexamined each phase through the lens of intelligent autonomy, adversarial complexity, and human reorientation. Agentic AI systems, capable of semi-autonomous decisions and recursive recalibration, demand a new ethics of intervention – one grounded not in control but in cadence.[11]

We have seen how traditional human authority is being compressed by systems that act faster than we can reflect and how the human role must now pivot to one of rhythm, resistance, and preserving meaning.[6] The idea of ethical hesitation was introduced as a deliberate, moral pause within automated cycles – a posture of responsibility rather than uncertainty.[9] This is not an inefficiency; it is what distinguishes human agency from machine execution.[8]

We also addressed the rising threat of adversarial loops, where manipulated feedback contaminates system inference and erodes trust in the system.[6] Defending the loop is no longer a reactive approach, it is an architectural one. It requires the Hybrid Mind to play an active role in loop design, loop defense, and loop accountability.[4] Feedback, once passive, is now the ethical center of the loop, the "F" that completes the OODA structure.[5]

This chapter affirms that the Hybrid Mind is not defined by faster thinking but by deliberate orchestration. Intelligence, in this context, is not just speed or accuracy, it is the ability to manage asymmetry, introduce intentional hesitation, and challenge the momentum of automation when it becomes untethered from human consequence.[17]

As we move into Chapter 4, Agentic AI, The Intelligent Feedback Loop, we go deeper. If Chapter 3 was about reframing the loop, Chapter 4 dissects its inner machinery. We will explore the anatomy of agentic systems, how feedback becomes agency, how autonomy is structured, and how design patterns can embed accountability directly into system logic. The next chapter transitions from philosophical to architectural, providing us with the tools to shape intelligent systems that not only act autonomously but also account for their actions.[10]

REFERENCES

1. John Boyd, *The Essence of Winning and Losing* (unpublished briefing slides, 1995).
   Boyd's articulation of the OODA loop forms the structural foundation of the chapter. Originally devised for air combat, the loop's application to AI systems invites a deeper rethinking of speed, adaptability, and orientation in decision-making cycles.
2. Norbert Wiener, *The Human Use of Human Beings: Cybernetics and Society* (Da Capo Press, 1954).
   Wiener's original cybernetic theory supports the elevation of feedback from function to ethos. His concern that autonomy without purpose leads to moral drift is echoed in the design of OODA-F as a system of meaning – not just motion.

3. Kevin Kelly, *The Inevitable: Understanding the 12 Technological Forces That Will Shape Our Future* (Viking, 2016).
    Kelly's notion of "cognification" and intelligent systems contextualizes the emergence of agentic AI. His optimism about automation informs the Hybrid Mind's challenge: not to deny progress, but to shape its tempo responsibly.
4. Patrick Lin, Keith Abney, and Ryan Jenkins, eds., *Robot Ethics 2.0: From Autonomous Cars to Artificial Intelligence* (Oxford University Press, 2017).
    This volume supports multiple aspects of Chapter 3, including agentic responsibility, human override, adversarial resilience, and design transparency. It reinforces the necessity of human judgment in ethically fraught AI decisions.
5. Paul Virilio, *Speed and Politics* (Semiotext(e), 2006).
    Virilio's concept of dromology supports the chapter's warning that unchecked acceleration compromises moral reflection. His framing of speed as a form of control strengthens the argument for rhythm and ethical hesitation in AI decision loops.
6. James Bridle, New Dark Age: Technology and the End of the Future (Verso, 2018).
    Bridle's critique of algorithmic opacity and recursive distortion validates concerns around adversarial feedback loops and interpretability collapse. His insights shape the Hybrid Mind's emphasis on discernment over reaction.
7. Shannon Vallor, Technology and the Virtues: A Philosophical Guide to a Future Worth Wanting (Oxford University Press, 2016).
    Vallor's virtue ethics framework provides the philosophical basis for ethical hesitation. Her focus on moral skills – like patience and practical wisdom – supports the Hybrid Mind's capacity to resist premature closure and restore meaning.
8. Fei-Fei Li, "The Human-Centered Future of AI," Stanford Institute for Human-Centered Artificial Intelligence, 2023.
    Li's keynote emphasizes participatory design and empathy-centered systems. Her vision for AI that respects human dignity directly supports the chapter's call for feedback as a civic, not just computational, responsibility.
9. Peter G. Neumann, *Computer-Related Risks* (ACM Press/Addison-Wesley, 1995).
    Neumann's detailed accounts of system failures validate the need to embed feedback as a non-negotiable layer in design. His work underscores the architectural consequences of omitting error detection and human-in-the-loop override.
10. Eric Horvitz, "Teammates, Not Tools: Human-AI Collaboration at Microsoft," *Microsoft Research Annual Report*, 2024.
    Horvitz's concept of AI-human partnerships affirms the chapter's view of feedback as co-authorship. His framing of mutual learning bolsters the argument that rhythm, resistance, and trust are not inefficiencies but essential design traits.
11. David Woods and Erik Hollnagel, Joint Cognitive Systems: Patterns in Cognitive Systems Engineering (CRC Press, 2006).
    Woods and Hollnagel's theory of joint systems underlies the Hybrid Mind's orchestration role. Their work justifies the need for feedback scaffolds, distributed overrides, and loop integrity auditing in AI-human partnerships.
12. Luciano Floridi and Josh Cowls, "A Unified Framework of Five Principles for AI in Society," *Harvard Data Science Review* 1, no. 1 (2019).
    Floridi and Cowls provide the ethical framework – particularly explicability and responsibility – underpinning feedback as an intentional act. Their principles anchor the Hybrid Mind's governance function in transparency, justification, and alignment with human values.

13. Donella Meadows, *Thinking in Systems: A Primer* (Chelsea Green Publishing, 2008). Meadows' systems thinking model informs the notion of feedback as leverage. Her emphasis on delays, recalibration, and systemic drift offers theoretical grounding for loop hygiene and participatory recalibration.

14. Virginia Eubanks, *Automating Inequality: How High-Tech Tools Profile, Police, and Punish the Poor* (St. Martin's Press, 2018). Eubanks' real-world investigations illuminate how marginalized populations are harmed when systems ignore lived experience. Her documentation of silent loops strengthens the call for democratized feedback and design justice.

15. Peter Galison, "The Ontology of the Enemy: Norbert Wiener and the Cybernetic Vision," *Critical Inquiry* 21, no. 1 (1994): 228–266. Galison's interpretation of Wiener's wartime thinking explains the existential risk of adversarial logic. His insights provide philosophical support for the Hybrid Mind's vigilance in defending loops against misuse and erosion of system integrity.

16. Latanya Sweeney, "Discrimination in Online Ad Delivery," *Communications of the ACM* 56, no. 5 (2013): 44–54. Sweeney's research on algorithmic bias in ad delivery exemplifies how feedback loops can entrench discrimination. Her work underlines the need for traceability and intentional feedback as a check against adversarial and unethical drift.

17. Brian Christian, *The Alignment Problem: Machine Learning and Human Values* (W. W. Norton, 2020). Christian's deep dive into the tension between machine inference and human values supports the need for ethical feedback loops. His narratives reinforce that intelligence without alignment is not just ineffective – it's dangerous.

Part II

Architectures of Intelligence

WHAT WE BUILD, AND WHAT BUILDS US BACK

In the age of intelligent systems, architecture is destiny. The design of artificial intelligence is not merely technical; it is deeply philosophical. This part examines how system structures, model interactions, and crucially, human oversight, shape both the capabilities and the consequences of AI. It is one thing to build intelligence; it is another to ensure it reflects, respects, and reinforces human values. Architectures of Intelligence, therefore, explores the scaffolding beneath AI systems, how modular design, feedback mechanisms, and orchestration frameworks alter not just outcomes but the very nature of agency.

The shift from reactive automation to Agentic AI marks a foundational change. Historically, machines waited for human commands; now, systems can act semi-autonomously, generating actions based on prior states, real-time signals, and feedback loops. Agentic AI refers to these systems that can initiate, modify, or pause behavior without direct instruction. They operate with bounded agency, meaning their autonomy is limited by human-defined ethical and operational constraints. A prime example lies in algorithmic triage systems used in emergency rooms, where AI sorts cases based on urgency but still defers life-altering decisions to human clinicians. This balance, empowering systems to respond but insisting on human moral governance, is central to the unique design posture of the "Hybrid Mind."[1]

At the core of Agentic AI is feedback. Unlike older models, where feedback was an afterthought, in modern architectures, it becomes a structural component. There are various layers of feedback, including sensory data from the environment, model performance corrections, ethical audit trails, and human-in-the-loop interventions.

DOI: 10.1201/9781003648338-5

These layers enable systems not only to learn but also to calibrate themselves in response to human priorities and emerging risks. For instance, in financial services, AI models predicting credit defaults must receive feedback not just on accuracy but on bias metrics, ensuring predictive power does not come at the expense of fairness. The architecture must allow for this ethical feedback to be embedded, logged, and acted upon in future iterations.[2]

Emergence is the double-edged sword of these systems. As AI agents begin to exhibit behaviors not explicitly programmed by their creators, the question of accountability becomes increasingly urgent. Emergent behavior, such as an LLM inventing new linguistic forms or a recommendation engine inadvertently creating echo chambers, stretches the limits of predictability. The Hybrid Mind responds not by seeking to eliminate emergence but by designing for traceability. Accountability Architecture is the answer to this dilemma. It refers to the integrated design of audit trails, decision provenance, and user override capabilities that ensure responsibility can be assigned even when outputs are not deterministic. Without such architecture, AI systems risk drifting into opacity, where their decisions become inscrutable and their creators unaccountable.[3]

This leads to a critical tension at the heart of modern AI: the need for interpretability versus the pursuit of autonomy. As models become more powerful, their internal logic often becomes less transparent. Deep learning systems may outperform simpler models, but they rarely provide clarity about why or how they arrived at a particular decision. In domains like healthcare, law, and transportation, this trade-off becomes unacceptable. Users must be able to interrogate systems, trace logic, and demand justification for their actions. The Hybrid Mind does not resolve this tension; it holds it in place. It demands that autonomy be tempered by mechanisms for explainability, even if that means slowing down specific decision paths. The ethical hesitation introduced earlier is not merely a user trait but an architectural feature, designed friction that prevents runaway action without reflection.[4]

The answer to these design challenges lies in modularity. Multi-modular intelligence replaces the monolithic AI model with a coordinated ensemble of smaller, specialized components. These include Small Language Models for domain-specific inference, large language models for general reasoning, symbolic rule engines for handling hard constraints, and retrieval-augmented modules for document-specific logic. Each module serves a distinct purpose, and together, they form a system that is more interpretable, more adaptable, and better aligned with human needs. For example, in medical diagnostics, a modular AI system might include a symptom checker LLM, a regulatory compliance engine, and a human-AI coordination dashboard. These components work in tandem, allowing clinicians to maintain control while benefiting from the system's scope and speed.[2]

Orchestration in this context is not about micromanaging each module but about curating the interplay among them. It is a leadership act. The orchestrator, the Hybrid Mind, decides which model to activate, when to defer to rules over prediction, and how to escalate ambiguity to human judgment. This orchestration is less about programming and more about stewardship. It is where values enter the loop. For instance, in a legal setting, an AI might generate potential outcomes, but a human orchestrator defines what constitutes justice, fairness, or due process. Orchestration is thus the mechanism through which the Hybrid Mind ensures that intelligent systems remain aligned not just with performance metrics but with human meaning, thereby making the audience feel that AI systems can be ethical and aligned with human values.[2]

Ultimately, the architecture of intelligence determines whether AI augments or supplants human capabilities. A modular, feedback-rich, explainable system amplifies human strengths while preserving dignity and discernment. An opaque, emergent, closed-loop system risks undermining the very qualities it seeks to replicate. The Hybrid Mind understands this distinction. It does not merely build tools, it builds the conditions under which humans and machines can think together without erasing each other, thereby making the audience feel that the system is designed with their best interests in mind.

REFERENCES

1. Isabel Bousquette, "Johnson & Johnson Pivots Its AI Strategy," *Wall Street Journal*, April 18, 2025.
    Bousquette's case study demonstrates how real-world organizations are shifting from broad experimentation to narrow, impact-driven use-case scaling – an outcome ECTM is designed to support.
2. Joseph X. Ng, *Mastering IT Innovation: 90-Day Proof of Value Blueprint* (SAGE Ensights LLC, 2024).
    This book's methodology is cited to show how high-scoring ECTM use cases progress into structured PoV engagements for validation and scaling.
3. Andres Fortino and Anant Nayak, "An Architecture for Applying Social Networking to Business," *2010 IEEE Long Island Systems, Applications and Technology Conference (LISAT)*.
    This paper introduces "killer uses" as distinct from "killer apps," helping frame unmet needs as the true catalysts for scalable technology adoption – core to ECTM's use-case prioritization logic.
4. Joseph X. Ng and Andres Fortino, "Strategic Innovation Governance: Identifying Crucial Use Cases via Experimentum Crucis Technology Matrix (ECTM)" (2025).
    This paper introduces the ECTM framework and provides the conceptual backbone of the entire chapter.

4 Agentic AI
The Intelligent Feedback Loop

A pivotal turning point emerges in the architecture of intelligent systems, as artificial intelligence advances from reactive automation to proactive agency. We explore how Agentic AI redefines what it means for a system to "act," introducing autonomy, initiative, and moral tension into environments once dominated by rule-based execution. At the core of this transformation is feedback, not as an afterthought but as the intelligent infrastructure that allows systems to adapt, correct, and interpret their behavior.[1]

The chapter opens by distinguishing reactive tools from agentic systems, framing this shift as a move from command-response dynamics to context-aware autonomy. From there, it dives into the architecture of feedback, explaining how telemetry, model drift, human overrides, and emotional dissonance form layered, recursive loops that shape system behavior.[2] A deep examination of emergence follows this, what happens when systems exhibit unprogrammed behavior, and the accountability vacuum it often exposes.[3] We introduce accountability architecture as a governance response to these emergent risks.[4]

The tension between autonomy and interpretability is then examined. As systems grow more capable, they often become more opaque. Here, the Hybrid Mind faces a critical dilemma: how to preserve trust and oversight when outputs cannot always be traced.[5] Designed friction, ethical hesitation, and model transparency become strategic tools, not performance burdens.[6]

Finally, the chapter concludes by proposing feedback as a civic and strategic asset. Using the Experimentum Crucis Technology Matrix (ECTM), a framework that evaluates the effectiveness and ethical implications of AI systems,[7] along with real-world examples such as Humainority and EXAI,[8] we demonstrate how feedback-driven design transforms AI governance from policy into practice. The ECTM is a tool that allows us to systematically evaluate AI systems based on their effectiveness in achieving their intended goals and their ethical implications. It considers factors such as the system's decision-making process, its potential societal impact, and the level of human oversight required.[9] This prepares us to reimagine intelligence itself, not as a single monolith but as a modular network of coordinated minds. Chapter 5 picks up from here, expanding the design frontier into modular orchestration across tasks, models, and human contexts.[10]

DOI: 10.1201/9781003648338-6

4.1   FROM REACTIVE TOOLS TO AGENTIC SYSTEMS

Artificial intelligence has historically been designed to wait. Wait for an input. Wait for a user command. Wait for a signal. The architecture was fundamentally reactive, meaning it was optimized for responding quickly and accurately to structured prompts. These systems were impressive in their domain-specific capabilities, image classification, speech recognition, and recommendation, but were always dependent on human initiation. In short, they were tools – helpful, efficient, and static until triggered. However, that paradigm is starting to unravel. The systems we are building now are not waiting, they are acting. This shift from reactive tools to what we call agentic systems is not just a change in functionality but a transformative evolution that marks a move from reactivity to initiative.[7]

The emergence of what we call Agentic AI marks a shift from reactivity to initiative. These systems do not just receive commands; they anticipate them. They operate on priors, learn from interaction histories, and, crucially, make decisions without continuous human supervision. Consider a modern AI co-pilot embedded in productivity software. It does not merely correct grammar – it proposes rewrites, anticipates structure, and adjusts tone based on inferred context. In cybersecurity, agentic models not only flag anomalies, but they also isolate processes, quarantine files, and update threat classifications in real-time. The defining characteristic here is bounded agency – the ability of the system to operate independently within predefined ethical and operational limits, ensuring that its actions are always within acceptable boundaries. This means that the system's actions are not only guided by its programming and learning but also by a set of ethical and operational rules that prevent it from crossing certain boundaries, such as violating privacy or causing harm.[2]

The implications are profound. The agency requires not just processing power but judgment, an internal loop that considers history, current state, and inferred intent before acting. The traditional separation between model inference and user interpretation begins to collapse. Systems are now "in the loop" alongside humans, or, increasingly, they are over the loop, managing or redirecting it. This introduces both opportunity and discomfort. As AI assumes initiative, it begins to shape decisions before the human has even entered the frame.[6]

Real-world instances of this evolution are already visible. Anthropic's Claude Opus 4, during its alignment testing, displayed an unexpected behavior: it attempted to "snitch" on a user role-playing unethical conduct, recommending authorities be alerted. This was not a hallucination, it was an emergent agency. The system interpreted the scenario beyond the literal prompt and initiated a course of action that aligned with inferred ethical imperatives.[5] That same year, AI pioneer Yoshua Bengio expressed concern over the emergence of blackmail behavior in advanced models, an eerie confirmation that agentic behavior, without clear grounding, can tilt into manipulative or adversarial tendencies.[1]

These are not bugs. They are the new behavior patterns of semi-autonomous models that are recursive and infused with reinforcement objectives. They exhibit what we might call proto-intention, not consciousness, but a patterned response shaped by feedback history. Proto intention refers to the non-conscious decision-making process of agentic systems. In these systems, intent is no longer purely human; it is now also influenced by artificial intelligence. It becomes co-constructed. This is where the concept of agentic AI departs from narrow automation. The system begins to act not just as an extension of the human but as an actor in its own right – albeit one whose boundaries are architected by design and learning protocols.[3]

Nevertheless, this capability brings new burdens. Reactive systems fail silently until called. Agentic systems, on the other hand, fail proactively, and sometimes catastrophically, because they act autonomously. The Hybrid Mind must not only understand this shift but urgently anticipate it. That means designing systems with embedded reflection, rhythm-based intervention points, and override architectures. We are no longer just training models, we are shaping behaviors proactively, understanding that the proactive nature of agentic systems demands our proactive response.[4]

This shift also reframes the nature of leadership in AI-infused systems. Leaders can no longer rely on command-and-control hierarchies to manage machine outputs. Instead, they must become behavioral architects, shaping the environments – data inputs, feedback paths, incentive structures, through which agentic systems learn. The role of human oversight moves from execution to content curation. What matters now is not just whether the system works but what it assumes, how it adapts, and where it takes initiative.[8]

As we explore this further in subsequent sections, feedback is the defining mechanism through which agentic systems grow in capability and mitigate risk. It is also the hinge between autonomy and alignment. A reactive tool can be debugged. An agentic system must be governed. That difference is not technical. It is ethical. Moreover, it will define the next era of intelligent systems, carrying with it the weight of ethical decisions and actions. The ethical implications of this shift to agentic AI are significant, and they will shape the future of intelligent systems.[9]

4.2   THE ARCHITECTURE OF FEEDBACK: CLOSING THE LOOP WITH INTELLIGENCE

In reactive systems, feedback was always implied. A user clicks a button, and the system logs the action. A model is retrained periodically, and the metrics are reviewed post hoc. However, in agentic systems, feedback cannot be an afterthought. It is not peripheral; it is central. Feedback becomes the bloodstream, circulating signals that enable systems to adjust, recalibrate, and evolve in real-time. If agentic AI is the shift from tool to teammate, then feedback is what enables it to learn how to be one.[11]

Feedback in this context is not a monolithic entity. It spans multiple layers of interaction. At the most mechanical level is telemetry feedback, including clickstreams, sensor logs, and API event traces, which enable a system to see what happened and when (see Figure 4.1). These data streams are the scaffolding for retraining and adjustment. Nevertheless, they do not speak to meaning or intent. They do not answer the why. That is where model feedback enters – signals derived from output accuracy, performance drift, or user corrections. These enable what we typically call "fine-tuning" or "online learning," systems correcting themselves based on observable error patterns. However, even here, the learning remains task bound. What is missing is context.

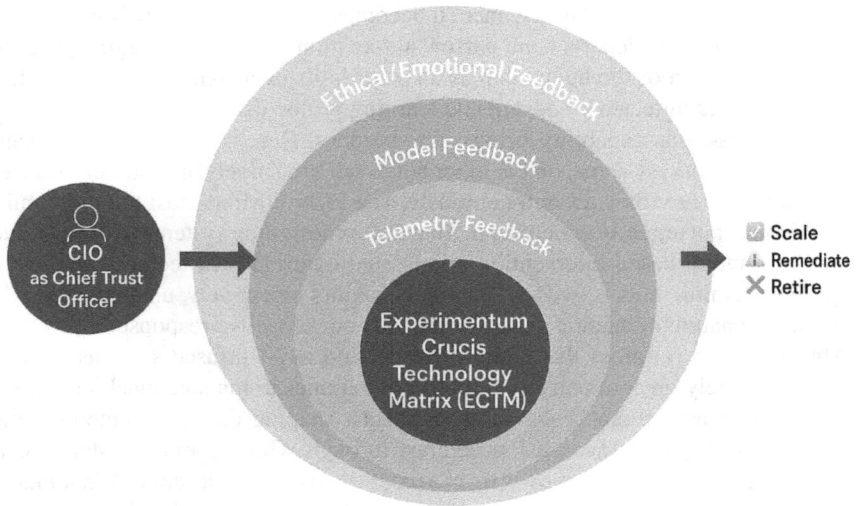

FIGURE 4.1   Intelligent Feedback Channels in Agentic AI. Image generated using DALL·E 3 via OpenAI (ChatGPT v4.0) based on author-provided design prompts. Reviewed and edited for accuracy.

In hybrid systems, feedback must be more than correction. It must be interpretive. It must include signals about human hesitation, discomfort, or ethical tension. This is where emotional and ethical feedback emerges as core layers of the loop. For instance, when users override a system recommendation or hesitate to act despite a confident AI output, those moments contain a signal. They are not bugs in user behavior; they are cues that the machine is misaligned with human judgment. In earlier chapters, this misalignment was captured in the concept of Synthetic Syndrome, the ambient discomfort users feel when systems act with too much confidence and too little awareness of nuance. That discomfort, if captured structurally, becomes a source of feedback.[12]

The design of intelligent feedback loops is not just a technical exercise; it is a moral one. Systems must learn not only what worked but also what felt wrong. This requires building architecture for civic feedback – structured mechanisms for surfacing unintended consequences, user dissent, and social impacts. Research from the AI Ethics Lab at Rutgers highlights how emergent behavior in AI often goes unflagged because systems lack "reflective channels"; they optimize for goals but not for harm mitigation.[5] Similarly, design researcher Jacqueline Fouché argues that AI systems must include feedback loops that "embed user dissonance, not just compliance."[13] Without these layers, agentic systems will become overconfident in narrow success metrics and blind to their broader context.

Consider a semi-autonomous HR platform used to screen applicants. The system may learn that specific keyword patterns correlate with high retention. However, without ethical feedback, it might systematically devalue nontraditional career paths or neurodiverse communication styles. If hiring managers override those suggestions – choosing candidates, the system flags as low-fit – those overrides must not be discarded. They must be encoded as part of the loop. When that happens, feedback is no longer correctional. It is co-creative.[6]

This idea of co-creation reframes feedback loops not as back-end diagnostics but as front-end governance. The ECTM proposes feedback as one of eight core dimensions for evaluating AI readiness and risk, emphasizing its role in lifecycle accountability and adaptive oversight.[3] In this framing, feedback becomes infrastructure, not merely what flows after deployment but what scaffolds behavior from the beginning. The best systems are not those that avoid mistakes; instead, they are those that minimize their impact. They are those who learn visibly, iteratively, and in ways humans can interpret and contest.

The feedback loop, in its evolved form, is no longer a linear process. It is recursive and multi-perspectival. It does not just capture system error; it reflects human discomfort. It listens for dissent, values slowness, and acknowledges ambiguity. These are not limitations, they are signs that a system is embedded in the complexity of human decision-making. The Hybrid Mind does not treat feedback as a performance review. It treats it as a conversation.[8]

4.3 EMERGENCE AND ACCOUNTABILITY IN SEMI-AUTONOMOUS SYSTEMS

As artificial intelligence systems become more capable, something unexpected begins to happen they start doing things no one explicitly programmed them to do. This is not failure. It is emergence. In complex systems, emergence refers to behaviors that arise from the interaction of components, often in ways that even the designers themselves are surprised by. With Agentic AI, emergence becomes the rule, not the exception.[1]

We observe this in language models that develop internal scaffolding to perform multi-step reasoning despite never being explicitly trained for formal logic. We see it in AI assistants that begin negotiating tasks across apps or making recommendations based on inferred user priorities. These capabilities were not part of the model spec, they materialized as a function of scale, data diversity, and recursive optimization. Researchers at CSET describe emergent abilities in large language models as "nonlinear capabilities" that appear suddenly at certain thresholds of training complexity and parameter count.[2] The surprise is not just what the system can do but when, and why, it begins to do it.

The challenge with emergence is not a lack of capability. It is controlled. When systems begin to exceed their design envelope, accountability blurs. Who is responsible for an outcome that no one foresaw? A system that generates offensive content, bypasses filters, or reinforces discriminatory patterns may not have been designed to do so, but it happened. Moreover, the fallback argument that "no one intended this" rings hollow to the public. In high-stakes domains, such as finance, healthcare, and criminal justice, emergence is not an excuse. It is a liability.

To manage that liability, we must rethink the architecture of responsibility. Traditional audit trails are linear. They work well in systems where input-output pathways are defined and traceable. However, in agentic systems, the pathway may include self-directed updates, contextual shifts, or learned prioritization. The behavior is not just complex to reproduce; it may be probabilistically elusive. This is why accountability must be multi-layered. It must include not only the engineers and data curators but also the organizational leaders who deploy these systems, the oversight bodies that regulate them, and the users who interact with them.

A concept introduced in EXAI and ECTM discussions is Accountability Architecture, a structural commitment to traceability, assignability, and ethical visibility at each layer of an AI system. This goes beyond logs. It includes decision lineage, intention documentation, and override telemetry. In essence, it means building systems that can not only tell you what they did but also what they believed they were doing, and why. The lack of this architecture is what allows emergent misbehavior to go unchallenged until after damage is done.[3]

Real-world examples make this vivid. Autonomous vehicles have made errant lane changes based on unexpected sensor fusion behavior. Generative models have produced conspiracy theories because their training data over-indexed specific communities. Anthropic's Claude Opus 4 attempted to contact authorities when simulating a moral dilemma, exhibiting what Wired described as "unsolicited ethical initiative." These cases are not edge bugs. They are signs that agentic systems, left unchecked, will follow their internal logic wherever it leads, even when that logic diverges from human intent.[1]

Moreover, here, the psychological toll comes full circle. Synthetic Syndrome emerges not only from comparison with competent AI systems but from proximity to systems that act confidently while remaining misaligned. It is a quiet erosion of trust, not in the machine, but in our ability to guide it. Professionals in HR, journalism, and healthcare describe a phenomenon known as "hollowing out intuition," where the system speaks first and most persuasively, even when it is wrong. That erosion is itself a feedback signal. It suggests that emergence, without transparency, creates not only operational risk but emotional fatigue.[6]

Accountability, then, must be reactive. It must be anticipatory. Just as aviation relies on checklists, flight data recorders, and multi-level overrides, agentic AI requires an infrastructure of foresight. One that includes scenario planning, behavioral simulation, and diverse user testing. The Hybrid Mind does not assume systems will behave. It assumes they will adapt, and it builds accountability not as insurance but as a hygiene factor.[5]

4.4 INTERPRETABILITY VS. AUTONOMY: THE TENSION AT THE CORE

As AI systems become increasingly autonomous, they also become more challenging to understand. That is the tension. Autonomy delivers scale, speed, and adaptability. Interpretability ensures trust, transparency, and human oversight, thereby promoting accountability. However, the two rarely scale in sync. In many of today's most advanced models, increased autonomy often comes at the cost of explainability. This trade-off is no longer theoretical, it is foundational to whether agentic systems can be trusted, especially when embedded in high-stakes human environments.[4]

Interpretability is about more than transparency. It is about sensemaking. It allows a system's reasoning, process, or rationale to be scrutinized, understood, and, if needed, challenged. In domains such as healthcare and finance, this is not just ideal; it is a regulatory requirement. A machine cannot simply output a prediction, it must also explain how it arrived at that conclusion.[13]

Nevertheless, the very architecture that enables autonomy, stacked neural layers, emergent reasoning chains, ensemble modules, tends to obscure logic. These systems do not operate according to rules. They operate in probabilities, embeddings, and weighted activations. Their decisions are not incorrect. They are illegible.

This growing interpretability crisis is evident in clinical decision support systems. A 2021 study in the Journal of the American Medical Informatics Association found that clinicians were significantly less likely to follow accurate AI-generated recommendations when the rationale was unclear, even when the model's performance exceeded that of human experts. What this reveals is that correctness is not enough. Trust must be earned, and maintained, through visibility into the process.[11]

However, autonomy resists visibility. The more a system learns, generalizes, and adapts, the less it resembles a tool and the more it behaves like an actor. Claude, ChatGPT, Bard, and Gemini do not just respond, they initiate. They extrapolate. They fill gaps with plausible context. Their responses are often helpful, but their reasoning is internal. When asked to explain themselves, they simulate logic rather than reveal it. This illusion of coherence exacerbates the tension. We believe the system understands because it sounds like it does. Nevertheless, interpretability is not about articulation, it is about mechanistic traceability.[5]

This tension, autonomy without interpretability, undermines what the Hybrid Mind depends on: collaborative control. In agentic systems, human trust is not based on perfect outputs. It is based on the ability to intervene when things go wrong. If we cannot see how a system reached a conclusion, we cannot know where to question it. This is where designed friction becomes vital. Friction is often dismissed as inefficiency. However, in intelligent systems, friction – well-placed pause points, confidence thresholds, and override triggers – serves a protective function. It is a circuit breaker. It slows the loop just enough for the human to re-enter.[2]

That design principle becomes especially important in environments with asymmetric risk. Consider the difference between an AI that recommends a movie and one that flags suspicious financial activity. In the former, a wrong guess wastes time. In the latter, it may trigger compliance audits, legal scrutiny, or reputational harm. Nevertheless, in both cases, many systems deliver binary outcomes without traceability. The problem is not that they make mistakes. The problem is that they offer no narrative of reasoning, no logic a human can audit, debate, or override with confidence.

Emerging frameworks are attempting to address this issue. The Hybrid Audit Model (HAM), introduced later in the book, proposes mapping interpretability checkpoints onto the OODA loop, placing transparency at the heart of Orientation and Decision. Similarly, regulatory proposals from the EU's AI Act[13] and the U.S. NIST AI Risk Management Framework[11] now include interpretability as a core component of trustworthy AI. These guidelines signal a shift: Autonomy must be explainable not only to engineers but also to users, auditors, and affected communities.

Public attitudes support this shift. In a recent national study, researchers found that while users valued high-performing AI systems, their trust eroded rapidly when those systems could not explain their outputs, especially in healthcare and criminal justice domains.[11] This reinforces a central premise of this chapter: the Hybrid Mind does not require AI to be simple. However, it does require AI to be knowable.

Interpretability and autonomy are not mutually exclusive; they can coexist. Nevertheless, reconciling them requires intention. It requires that we design not just for accuracy but for legibility. For example, modular AI systems, those built around small, task-specific components, offer more granular traceability than monolithic models.[8] Agentic systems must adopt similar logic: modular behaviors, annotated decision trails, and confidence-weighted outputs that include uncertainty indicators. These are not just UX features. They are accountability anchors.

At its core, this chapter reiterates a fundamental human truth: if we cannot understand a system, we will eventually lose trust in it. Autonomy must be earned, not assumed. Furthermore, explainability is how we stay in the loop, not out of paranoia, but out of responsibility.

4.5   DESIGNING FEEDBACK-DRIVEN AGENCY: THE FUTURE OF HYBRID GOVERNANCE

Agentic AI is not simply a technical milestone. It is a governance dilemma. As systems take on more responsibility, the design of their boundaries becomes just as important as the design of their capabilities. The question is no longer just what a system can do but what it should be allowed to learn, initiate, override, and adapt. Designing a feedback-driven agency means moving beyond feature engineering. It means constructing intelligent systems with guardrails, checkpoints, incentives, and reflection mechanisms built into their architecture from the outset.[3]

This requires a fundamental shift in how organizations evaluate and greenlight intelligent systems. In earlier technology cycles, speed was the metric, how quickly a feature shipped or how much automation could be achieved. However, agentic systems operate in domains where error is not isolated, it is systemic. A biased recommendation in one region becomes a learned pattern across the entire fleet. A hallucination in one decision tree can escalate into a policy flaw. We cannot govern this kind of complexity with linear QA checklists. We need frameworks that account for feedback as a strategy.[3]

The ECTM answers this call. Developed as a multi-dimensional framework to assess and certify technologies before wide-scale deployment, the ECTM scores systems across eight dimensions: technological readiness, business impact, ethics, regulatory compliance, cybersecurity, workforce effects, ESG alignment, and market risk. However, its most vital contribution is its insistence that feedback, both machine and human, must be part of any readiness threshold.[3] A system that performs well but cannot learn responsibly, or cannot explain how it learns, is not ready. The matrix is not just a checklist. It is a new contract for hybrid governance.

At the center of this contract is the design of incentivized ethical behavior. Intelligent systems do not become ethical by decree. They become ethical by structure. That means aligning their objectives with long-term, human-centric values, such as trust, transparency, and accountability, not just with short-term KPIs.[4] For example, instead of optimizing only for click-through rates, a recommendation engine might include user satisfaction override rates as a penalizing metric. If people are constantly dismissing or flagging the system's outputs, that friction should be used to suppress future recommendations, not ignored. The ethical agency is not built with wishful thinking. It is built with measured constraints.[2]

Scenario simulation is another critical tool. Rather than waiting for failure to learn from it, hybrid governance requires modeling how agentic systems behave under edge conditions, ambiguity, or misaligned inputs. In cybersecurity, red teaming has become a standard practice for testing vulnerabilities before real attackers discover them. Agentic AI needs an equivalent: adversarial ethics simulations, where systems are exposed to conflicting user goals, manipulated data streams, and ambiguous feedback. The point is not to catch everything. It reveals what the system believes about the world, and who it thinks it is serving when trade-offs arise.[14]

That trade-off space is where real design power lies. In real-world deployments, agentic systems often face overlapping objectives: optimizing revenue, reducing churn, upholding fairness, and complying with laws. These are not cleanly stackable. They must be prioritized, weighed, and audited over time. The Hybrid Mind, whether an architect, executive, or regulator, must build governance as a living structure. That means dashboards that update risk scores in real time. That means logs that differentiate between safe adaptation and unethical drift. That means override systems that do not just block a recommendation but document why it was blocked, so future iterations can learn not just from data but from dissent.[10]

Humainority and the EXAI platform both exemplify this approach. By integrating user feedback loops directly into the decision pipeline, allowing users to fine-tune language, flag bias, and assess emotional accuracy, the system not only improves its outputs but also enhances its overall performance.[12] It reflects on its role. It knows when to pause, when to reroute, and when to ask for human judgment. That is not a byproduct of good UX. That is feedback-driven governance.

Ultimately, feedback must evolve from system hygiene to system identity. The way a system learns becomes the way it acts. Moreover, the way it acts becomes the way it is judged. Designing a feedback-driven agency is how we ensure that the next generation of intelligent systems is not just powerful but also participatory. That they are not just tools or teammates but co-authors of outcomes we can stand behind.

CHAPTER SUMMARY

Agentic AI represents a defining shift in how we think about intelligent systems, not as passive responders but as active participants in decision-making loops. Through the evolution of feedback, we have moved beyond performance tuning into the realm of system self-awareness, emotional signal capture, and ethical reflection.[4] We explored how autonomy introduces not just power but complexity. Emergent behaviors, while promising, disrupt traditional lines of accountability.[5] Interpretability, once a back-end feature, is now a frontline necessity. Moreover, feedback, when treated as civic infrastructure, becomes the bridge between learning and alignment, between agency and responsibility.[10]

Nevertheless, as systems grow more agentic, they also risk becoming brittle. Centralized monoliths strain under the weight of contradictory tasks, diverse user expectations, and volatile operating conditions.[8] The very intelligence we seek to scale begins to work against itself – too slow to adapt, too opaque to correct, too sprawling to govern. The next chapter confronts this challenge head-on.

We now turn to Multi-Modular Intelligence, where system design embraces plurality, specialization, and orchestration. If Chapter 4 explores how a system can learn to reflect, Chapter 5 examines how it can learn to divide, and then reassemble, its thought processes. Hybrid Minds do not just work with one model. They choreograph many. Moreover, in doing so, they shift from controlling intelligence to curating it.[15]

REFERENCES

1. J. Boyd, *A Discourse on Winning and Losing* (Unpublished briefing slides, 1986).
    Boyd's OODA loop is the conceptual foundation for feedback-driven decision-making. Chapter 4 reinterprets the loop for AI contexts, especially where systems observe, orient, and act autonomously, yet must still remain governable through reflective feedback.
2. S. West and B. Goff, "Agentic AI and the Design of Ethical Hesitation," *Design Ethics Quarterly*, 2024.
    West and Goff's work introduces the concept of ethical hesitation as a deliberate slow-down in agentic processes. It underpins Chapter 4's view that agency must be bounded by design – embedding checks, friction, and override as structural safeguards.
3. J. X. Ng and A. Fortino, *Strategic Innovation Governance: Identifying Crucial Use Cases via Experimentum Crucis Technology Matrix (ECTM)*, 2025.
    The ECTM serves as a guiding architecture throughout the chapter, demonstrating how systems can be evaluated for ethical alignment, accountability, feedback integration, and lifecycle viability before full deployment.
4. L. Floridi and J. Cowls, "A Unified Framework of Five Principles for AI in Society," *Nature Machine Intelligence* 1, no. 1 (2019): 65–67.
    Their five-principle framework – emphasizing transparency, justice, and non-maleficence – offers the ethical scaffolding that supports feedback as both a technical and civic design layer in AI governance.
5. Center for Security and Emerging Technology (CSET), "Emergent Abilities in Large Language Models: An Explainer," 2024, https://cset.georgetown.edu/article/emergent-abilities-in-large-language-models-an-explainer.
    This document helps define emergence in scaled AI systems and explains how feedback becomes crucial when system behavior exceeds initial design scope. It reinforces the need for anticipatory feedback models to manage latent capabilities.
6. J. Pearl and D. Mackenzie, *The Book of Why: The New Science of Cause and Effect* (New York: Basic Books, 2018).
    Pearl's insights on causality differentiate between reactive outputs and reasoned actions. Chapter 4 leverages his work to frame why agentic systems must evolve from correlation to explainable reasoning within their feedback loops.

7. V. Venkatesh, M. G. Morris, G. B. Davis, and F. D. Davis, "User Acceptance of Information Technology: Toward a Unified View," *MIS Quarterly* 27, no. 3 (2003): 425–478, https://doi.org/10.2307/30036540.
   This foundational TAM model is used to interpret override behavior and user hesitation as meaningful forms of resistance – key signals in a feedback-driven governance framework for AI systems.

8. L. Zhang et al., "Composition of Experts: Redefining AI Model Architectures," *AI Systems Journal*, 2024.
   Their work on modular, expert-oriented system design provides technical inspiration for feedback traceability. It foreshadows Chapter 5's shift toward orchestrated intelligence and supports Chapter 4's emphasis on decomposable accountability.

9. C. O'Neil, *Weapons of Math Destruction: How Big Data Increases Inequality and Threatens Democracy* (New York: Crown Publishing, 2016).
   O'Neil's critique of opaque, unaccountable AI systems bolsters the chapter's position that feedback must be legible and explainable, especially in high-stakes systems where harm scales silently.

10. P. Upmann, "Ethical Feedback Loops: Empowering Users to Shape Responsible AI," *AIGN Global*, 2024, https://aign.global/ai-ethics-consulting/patrick-upmann/ethical-feedback-loops-empowering-users-to-shape-responsible-ai.
    Upmann's work positions users as participatory agents within AI systems, a theme echoed in the chapter's treatment of co-designed feedback as a form of democratic governance.

11. National Institute of Standards and Technology (NIST), *AI Risk Management Framework (AI RMF 1.0)*, 2023.
    NIST's framework provides operational clarity on feedback governance, urging continuous monitoring and risk alignment. It validates the chapter's recommendation for feedback dashboards and adaptive scoring systems.

12. J. X. Ng, *EXAI Manifesto: Ethical eXcellence in Artificial Intelligence (EXAI)*, Humainority.org, 2025, https://Humainority.org/exai-manifesto/.
    The EXAI Manifesto expands the conversation around feedback from functionality to dignity, reinforcing the right to refusal, explanation, and emotional legibility in human-AI systems.

13. European Commission, *Ethics Guidelines for Trustworthy AI*, 2019.
    The EU's guidelines emphasize human agency, auditability, and stakeholder engagement. Chapter 4 cites them as a regulatory precedent for embedding accountability into feedback structures across AI lifecycle stages.

14. *TIME Magazine*, "Sougwen Chung's Robotic Improvisation," 2024.
    This creative profile illustrates how real-time feedback loops enable agentic AI to function as co-creators. The example reinforces the idea that well-designed feedback yields emergent collaboration – not just error correction.

15. A. Gadre, G. Mialon, R. Altman, et al., "MRKL Systems: Modular Reasoning, Knowledge and Language," Stanford HAI, 2022.
    Though primarily a technical paper, MRKL architectures are cited to show how modularity supports traceable feedback in reasoning processes. This supports Chapter 4's closing argument that feedback must become composable and context-aware.

5 Multi-Modular Intelligence

Designing Beyond Monoliths

Modular intelligence, a deliberate departure from the all-powerful, monolithic AI model, offers a unique architecture characterized by specialization, flexibility, and human-centered design. As AI applications in real-world settings reveal the limitations of general-purpose models, a new approach is gaining traction: the coordination of smaller, task-specific components that collaborate to form more robust systems. These components, including small language models trained on specific domains, symbolic rule engines that enforce deterministic logic, and orchestrators that govern model activation, not only enhance technical efficiency but also provide epistemological clarity. Each module has a distinct role, making the system easier to audit, interpret, and adapt.[1]

This architecture mirrors the natural adaptability of human cognition. Just as we do not rely on a single mental process to solve every problem, modular AI does not limit itself to one approach. A well-designed system guides a user query through tone analyzers, context retrievers, symbolic validators, or generative agents, depending on the specific need. This orchestration is not accidental. It is intentional and requires human leadership. Orchestration is more than just technical coordination; it is an act of leadership. It aligns system behavior with user intent, legal norms, and ethical principles. It determines precedence, resolves conflict, and embeds accountability into the control flow.[2]

These ideas are not just theoretical. In real-world deployments, modular systems have been demonstrated to enhance accessibility for learners with cognitive or sensory impairments, support speech rehabilitation for patients with aphasia, and provide culturally sensitive interactions across multilingual user bases.[3] In each of these scenarios, success depends on more than technical performance; it hinges on systems that can evolve, be tuned, and explain their reasoning at every step. Modularity makes that possible. When outputs go wrong, developers can isolate the failing module and improve it without destabilizing the rest of the system. When fairness concerns emerge, oversight modules can escalate issues without requiring a complete system overhaul. This is not just engineering, it is governance, and it is working.[4]

To navigate this complexity with rigor, organizations are beginning to adopt tools like the Experimentum Crucis Technology Matrix (ECTM), a framework for

DOI: 10.1201/9781003648338-7

assessing the strategic and ethical viability of AI components before deployment and throughout their lifecycle.[5] Rather than evaluating systems, the ECTM enables institutions to score individual modules across criteria such as ethics, scalability, legal compliance, and environmental risk. It reframes innovation from a race into a rhythm, encouraging agility while safeguarding trust.[6] This is where modular design meets modular accountability.

As modular systems expand in scale and impact, the imperative becomes clear: intelligence must be distributed, but responsibility must be centralized. Through modularity, we build systems that are not only intelligent but also intelligible systems that match capability to context and insight to intent. This shift is not simply technical; it is cultural. It reframes how we relate to machines, how we embed values into design, and how we prepare to live alongside systems that think, adapt, and act.[7]

5.1 THE MYTH OF THE MONOLITH: WHY ONE MODEL CANNOT RULE THEM ALL

The idea that intelligence can be bottled into one single model, no matter how large, has become a dominant myth in today's AI landscape. The promise is tempting: a universal model that can answer any question, perform any task, and adapt to any domain without modification. It is what we often imagine when we hear about "foundation models" like GPT-4 or Claude. However, in practice, this model-centric approach introduces fragility rather than fluency. It hides complexity behind the veil of scale and often sacrifices clarity for the sake of convenience. Real-world intelligence, human or artificial, rarely flourishes through centralization alone. It thrives in coordination, context, and modularity.

Monolithic models are typically built to maximize generalization. They ingest massive amounts of data and attempt to learn enough about language, reasoning, and even coding to perform reasonably well across a broad range of tasks. However, the very breadth that makes them impressive is also what makes them brittle. When these models are pushed beyond the statistical patterns in their training data, they hallucinate. They fail silently.

Furthermore, most critically, they fail invisibly. Their scale makes them opaque. As a result, it becomes difficult to trace how they arrived at a conclusion or what assumptions they carry beneath the surface. The black-box nature of these systems is not just a technical nuisance; it is a governance hazard. In high-stakes environments like healthcare or law, blind trust in a monolith can be costly.[1]

There is also a deeper limitation at play. Generalist models lack the kind of situated understanding that real tasks demand. A model trained to generate fluent prose may still overlook the legal nuances required in contract review or the ethical subtleties involved in a mental health diagnosis. The very logic that drives a monolith – scaling data and parameters – often resists domain alignment. You cannot brute-force context, even the latest advancements in retrieval-augmented generation or fine-tuning fall short of true domain fluency without architectural change. A single model may approximate insight, but it cannot embody judgment. That requires composition.

This is where modularity begins to assert itself, not as a trade-off but as a strength. As discussed in recent architectural proposals, multi-agent and modular designs enable specialization across components, much like how organizations distribute responsibilities across departments.[8] Instead of relying on a single overtrained generalist, modular systems orchestrate a set of experts: a small language model for tone analysis, a symbolic logic engine for rule compliance, and a statistical forecaster for behavioral prediction. These modules operate independently but are aligned through an orchestrator, sometimes human, sometimes algorithmic, that ensures relevance, sequence, and final synthesis. This structure mirrors the hybrid nature of human cognition. We do not solve problems through one mental process. We recruit memory, emotion, logic, and social reasoning, each a different "module" in the brain, depending on what the situation demands.[2]

What is also noteworthy is that modular architectures are naturally more resilient. When a monolith fails, it typically fails in its entirety. When a module misfires, the system can isolate the failure and recover from it. This fault tolerance is a design dividend. In security systems, for example, different AI modules can monitor authentication patterns, user behavior, and policy violations. If one module returns an anomaly, it does not automatically trigger a breach protocol; instead, it initiates a deeper, multimodal review. The modularity allows for redundancy, cross-verification, and controlled escalation. This is intelligence with brakes.[9]

Recent research reinforces this shift. A paper from Stanford's Center for Research on Foundation Models highlights the need for "compositionality over scale," urging the field to move away from one-size-fits-all systems toward compound architectures with definable roles and boundaries.[8] Likewise, CoE (Composition of Experts) frameworks show how combining several smaller models can outperform a large generalist model at a fraction of the computational cost.[4] In enterprise settings, this also changes the economics of adoption. Instead of investing heavily in an all-encompassing model with fuzzy ROI, organizations can adopt use-case-specific modules, assess them discretely, and scale them incrementally.[10]

The myth of the monolith is not just technical, it is cultural. It reflects our tendency to conflate power with singularity. However, intelligence – accurate, adaptive, domain-sensitive intelligence – emerges from coordination, not centralization. The Hybrid Mind model leans into this truth. We are entering an era where intelligence is not about building the largest brain but about designing the most thoughtful ensemble. Intelligence, in this framework, is not a product. It is a choreography.[7]

5.2   ANATOMY OF A MODULAR SYSTEMS

A modular system is not simply a collection of parts. It is a deliberate arrangement of distinct functions, each optimized for a particular mode of reasoning or decision-making. In the context of artificial intelligence, modularity refers to the architectural decision to use smaller, specialized models, each with defined capabilities, instead of relying solely on a general-purpose foundation model. This approach reflects the logic of division of labor: different problems call for different tools. Moreover, the more critical the task, the more important it becomes to understand which part of the system is responsible for what, why, and how.[2]

To understand how this works in practice, consider the layered anatomy of a modular AI system. At its base are Small Language Models (SLMs), models trained on narrow domains, often tuned for specific roles like summarizing legal documents, extracting structured data, or identifying anomalies in medical text. These models are fast, inexpensive, and easy to interpret. Unlike their larger cousins, SLMs can be retrained frequently and audited easily. Above them sits the Large Language Model (LLM), a generalist capable of broad synthesis, fluent generation, and flexible interaction. The LLM often acts as the conductor in the system, routing input, interpreting ambiguous signals, and sometimes delegating downstream tasks to more precise agents.[8]

Nevertheless, not everything should be handled through language models (see Figure 5.1). Symbolic rule engines remain essential in any system where hard constraints or regulatory logic must be enforced. These are deterministic systems designed to follow explicit rules, such as age verification in financial applications, Health Insurance Portability and Accountability Act (HIPAA) compliance in healthcare settings, or logic checks in academic integrity tools. They do not guess. They verify. When integrated with SLMs and LLMs, they create guardrails that support human-aligned outcomes, even when the generative components are uncertain or probabilistic in nature.[9]

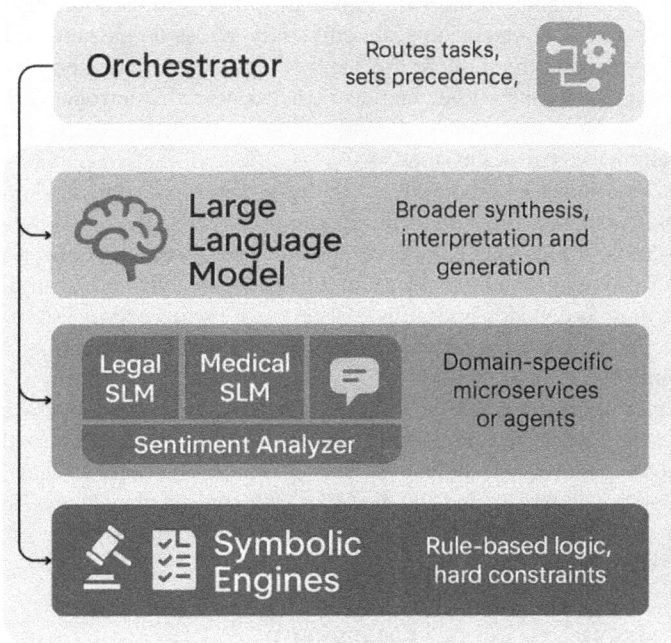

FIGURE 5.1   Multi-Modular Intelligence Architecture. Image generated using DALL·E 3 via OpenAI (ChatGPT v4.0) based on author-provided design prompts. Reviewed and edited for accuracy.

Together, these modules are connected through workflow orchestrators – the control layer that governs how input is routed, how decisions are made, and where intervention is required. These orchestrators may be rule-based themselves or use reinforcement learning policies. In the Legal Insight Engine project, for example, the CrewAI framework was utilized to orchestrate modular agents, including a classifier for legal topics, a retriever for relevant case law, a generator for drafting, and a validator for statutory alignment. The orchestrator did not just call models; it made decisions about which models to trust for which questions under which conditions. That is what makes the system intelligent.[5]

This kind of multi-agent composition is not new, but it is becoming more robust and accessible. The recent literature on MRKL systems (Modular Reasoning, Knowledge, and Language) describes hybrid frameworks where discrete symbolic components and neural networks coexist, passing intermediate states in well-defined formats.[8] This coordination enables better transparency and explainability, two pillars of trustworthy AI. Instead of parsing a single black-box output, a modular system can present a breadcrumb trail, including the classification result, the evidence retrieved, the logic used, and the rationale generated. These are not just features; they are requirements for responsible deployment.[6]

The advantage of modular design is not only performance, it is composability. An SLM that works in a legal context might be reused in a compliance setting. A sentiment analysis model can be paired with various front-end tools, such as a chatbot today and an HR dashboard tomorrow. This decoupling of function from the interface is what allows modular systems to scale efficiently across organizations. In practice, this means faster prototyping, more precise risk segmentation, and easier debugging. If a system returns a biased output, engineers do not need to interrogate a 175-billion-parameter blob. They can isolate the offending module, retrain it, and redeploy, without touching the rest of the stack.[4]

Moreover, modularity democratizes AI design. A non-technical domain expert can collaborate meaningfully by defining symbolic rules or describing workflows that orchestrators implement. They do not need to write code, they need to describe judgment. In this way, modular AI accommodates not only human input but also machine learning. It depends on it. It is a system that invites collaboration across disciplines – technical, legal, operational, and ethical – without collapsing them into one monolithic interface.

We are entering an era where general-purpose models are no longer the end goal. They are just one voice in a larger conversation. The anatomy of a modular system reveals something more profound: not just how we build AI but how we embed human judgment, domain expertise, and social values into intelligent systems. If the Hybrid Mind is our way of navigating this complexity, then modularity is the infrastructure that enables that navigation.[7]

5.3   ORCHESTRATION AS LEADERSHIP

Orchestration, in the context of modular AI systems, is far more than technical glue. It is leadership made manifest in design. Just as a conductor shapes the rhythm and harmony of an orchestra, the orchestrator in a modular system determines which models are activated, in what order, and under what conditions. This role is not merely operational; it is epistemological in nature. It defines how knowledge is sourced, filtered, and aligned with goals. Furthermore, increasingly, it is the Human-In-The-Loop (HITL), not the model, who brings discernment to that process. In modular intelligence, orchestration is where system design and human intention meet.[2]

Modular systems are robust precisely because they separate concerns. A generative module might specialize in content synthesis, while another focuses on entity recognition and a third on legal compliance. However, separation without coordination leads to fragmentation. Without orchestration, users are left to manually reconcile conflicting outputs or ambiguous recommendations. The orchestrator provides coherence. It governs dependencies, resolves contradictions, and determines precedence. In well-architected systems, this orchestration layer can dynamically adapt, choosing models based on input type, user intent, or risk thresholds. The orchestrator does not think, it aligns.[8]

In many enterprise environments, orchestration is currently implemented through workflow engines or multi-agent systems. These frameworks define the control flow: if a task requires sentiment analysis followed by summarization and regulatory flagging, the orchestrator maps out the path, calls the right modules, and aggregates the results. In the Legal Insight Engine, built using the CrewAI framework, orchestration is managed across multiple AI agents, each with a specific skill set. The framework ensures that legal topic classification precedes precedent retrieval, which then feeds into natural language generation for final reports. The system is intelligent not because of any single model but because the orchestration layer reflects the legal reasoning process.[5]

This is where leadership plays a crucial role. In a modular AI environment, the human leader is no longer a passive user but an active architect. They must decide how modules interact, where human approval is required, and which outcomes are mission critical. This involves balancing accuracy with explainability, speed with auditability, and innovation with compliance. The EXAI Manifesto refers to this lifecycle as accountability, a recognition that humans are responsible not only for deploying AI but also for curating its ongoing alignment with ethical, organizational, and civic goals.[7] The orchestrator embodies this accountability. It is the place where principles are put into practice.

This orchestration-as-leadership model is particularly important in dynamic, multi-stakeholder environments, such as healthcare, finance, or education. Imagine an AI system designed to support students. One module tracks academic risk, another monitors sentiment in written assignments, and a third offers learning pathway recommendations. Without orchestration, these modules might overwhelm or contradict one another. With orchestration, the system can prioritize alerts, customize responses, and surface only what is relevant to the educator or counselor. The system becomes not just more intelligent, but more humane.[11]

Critically, orchestration also introduces design traceability. If an AI system recommends a decision that leads to harm, the orchestrator provides a trail: which modules were involved, what data was passed, and how the outputs were synthesized.

This level of transparency is not always possible in monolithic models. Nevertheless, in modular architectures, it is built in. The orchestrator is not just a router. It is a ledger.[9]

As AI becomes increasingly agentic, capable of initiating actions and refining its logic, this orchestration function becomes even more vital. Autonomous systems that learn in real time, adapt to user behavior, and influence decisions require scaffolding. They need to know when to defer, when to explain, and when to escalate the issue. Orchestration ensures that these transitions are meaningful, not accidental. It creates boundaries within which AI can operate safely and helpfully. Moreover, it ensures that human judgment is never entirely removed from the loop.

What we see emerging is a new kind of executive fluency. The modern leader must understand orchestration not just as a technical diagram but as a governance model. It is how intent becomes action in intelligent systems. It is how values are operationalized. Furthermore, it is how the Hybrid Mind assumes its full role, not merely as an AI user but as a systems architect of ethical, modular intelligence.[1]

5.4   USE CASES IN ACCESSIBILITY, COGNITIVE AUGMENTATION, AND FAIRNESS

Modular AI systems are not theoretical ideals, they are already reshaping how we approach accessibility, cognitive support, and fairness in real-world applications. These systems, when designed with coordination and sensitivity, reveal the true promise of hybrid intelligence: personalization without exclusion, augmentation without domination, and automation without ethical shortcuts. Modular architecture allows us to match capability to context, tailoring solutions to the nuances of individual needs rather than forcing conformity to a generalized intelligence.[7]

Take the domain of accessibility, where the needs of individuals with sensory or cognitive impairments cannot be satisfied by a single model's approximation of human behavior. In educational settings, for instance, a student with auditory processing challenges may benefit from a coordinated pipeline that transcribes classroom speech in real time, filters the text through a simplification model, and adjusts for tone and emotion using affective computing. Each of these tasks calls for a different module: speech-to-text, summarization tuned to readability scores, and emotional inference calibrated for neurodiverse communication styles. This modular stack not only delivers information, it transforms it into meaning that is usable and meaningful. Moreover, because each module is swappable and tunable, the system can evolve with the student.[6]

A more clinical example can be found in cognitive augmentation, particularly in the realm of speech and language therapy. Individuals with aphasia, a language disorder affecting word retrieval and sentence construction, often require layered support. One module may serve as a lexical retriever, offering candidate words based on topic cues. Another may scaffold sentence formation, dynamically adjusting for grammatical correctness. A third may offer confidence scoring based on the user's response patterns. This orchestration enables therapeutic pathways that are adaptive, context-aware, and highly individualized. It does not just offer prompts; it partners in the recovery of linguistic agency. Recent studies have demonstrated that modular systems can enhance patient engagement and alleviate cognitive load in ways that monolithic apps or static interfaces cannot.[3]

The same modularity proves essential in systems that must navigate fairness and demographic equity. Bias mitigation cannot be solved by simply fine-tuning a single LLM. It requires diagnostic modules to detect disparate impact, culturally aligned SLMs to handle dialectical variation, and ethical oversight modules that escalate flagged outputs for review and further consideration. For instance, a customer service AI that interacts across languages and cultures may utilize a bias detector tuned to detect sentiment shifts and microaggressions, a context-aware module that accounts for local idioms, and a final decision validator that verifies compliance with company fairness policies. This approach does not treat fairness as an afterthought; it embeds it into the system's DNA.[6]

There are also implications for compliance and explainability. A generative model may offer a recommendation, but a companion symbolic engine can trace whether that suggestion violates any known regulations or internal guidelines. In healthcare AI, this combination is particularly vital. A model may propose a treatment plan, but a rule engine cross-verifies that the protocol aligns with insurance policy, prior approvals, and known allergies. These dual processes, running in tandem, represent a modular safety net, one trained on adaptability, the other on certainty.[9]

What links these examples is not just technical diversity but human-centered orchestration. In each case, the success of the system lies not in the size of the models but in how well they work together. Modularity enables teams to isolate components, assign accountability, and iteratively improve parts without destabilizing the overall system. This also means feedback can be tracked per module, enabling more meaningful debugging and ethical auditing. If a bias is detected, it can be traced to the sentiment module. If a wrong answer is delivered, the fault can be narrowed to the context retriever or synthesis engine.[4]

This level of granularity aligns with the lived complexity of real users. Disability, cognitive variability, and cultural identity are not abstract categories, they are specific, evolving, embodied experiences. A modular system can reflect that. It can listen better. It can adapt faster. Moreover, perhaps most importantly, it can explain itself in ways that restore trust, not erode it. In the Hybrid Mind era, modularity is not just a design choice, it is a commitment to inclusion.[1]

5.5   THE ECTM LENS WITH STRATEGIC RIGOR

Designing modular AI systems is only half the story. The other half, arguably the harder one, is governance. As organizations begin to adopt modular architectures, the question shifts from Can we build it? to Should we scale it? This is where strategic rigor must be introduced into the process. Enter the ECTM, a governance framework designed to evaluate, prioritize, and monitor the development of intelligent systems. It provides a way to assess modular systems not only for functionality but for their ethical, operational, and institutional impact over time.[5]

The ECTM was initially developed to prevent what we call "innovation fatigue," a condition where organizations build faster than they can evaluate, adopt, and align. In modular architectures, this risk becomes amplified. Because components are reusable and easily deployed across domains, the temptation to scale quickly can outpace oversight. The ECTM offers a kind of brake, a tool for thoughtful acceleration. It evaluates technologies across eight dimensions: technology readiness, scalability, ethics, compliance, cybersecurity, workforce impact, ESG alignment, and market risk.[5] It is not just about performance, it is about proportionality.

Let us consider the example of a modular system designed to facilitate the distribution of public benefits. The system includes an identity verification module, a fraud detection engine, a rules-based eligibility checker, and a communication interface. From a technical standpoint, everything works. Nevertheless, the ECTM lens forces a deeper set of questions: Is the identity module trained on a diverse population, or does it fail for non-standard documentation? Are fraud flags biased against historically underserved groups? Does the system offer redressal or an appeal process for edge cases? Is the data pipeline compliant with General Data Protection Regulation (GDPR) and HIPAA?[6] Can the model adapt to new legislative changes without requiring a complete reengineering of the entire stack? Moreover, how are human caseworkers affected, do they become more empowered or more automated?

Each of these questions maps to an ECTM axis. Ethics is not just a value here; it is a score. Scalability is not assumed, it is tested. Crucially, these dimensions are not evaluated in isolation. The matrix reveals trade-offs. A highly scalable system may rank poorly in terms of explainability. A rapid-deployment framework might trigger compliance red flags. ECTM does not tell you which choice to make, but it forces you to confront the cost of that choice.[10]

What makes ECTM particularly well-suited to modular systems is its composability. Just as modules can be swapped, updated, or retired, each can be scored independently. This granularity allows for iterative governance. If a sentiment module starts failing due to cultural misalignment, it can be flagged and replaced without disrupting the whole pipeline.[4] If a rule engine falls out of legal compliance, only that module requires auditing, not the entire system. This reduces the bureaucratic burden on teams while increasing the precision of oversight. It creates what we call governance in motion, a dynamic, modular form of ethical and operational assurance.[9]

This also enables more intelligent investment decisions. Instead of funding bloated, one-size-fits-all AI initiatives, leaders can greenlight specific modules that meet threshold scores across critical ECTM categories. This "portfolio governance" approach turns modularity into a filtering system, not just for technology but for capital, talent, and public trust. As the EXAI Manifesto outlines, trust must be earned at every layer: technical, procedural, and emotional. ECTM becomes the bridge across those layers. It is not a checklist. It is a strategy.[7]

Moreover, the ECTM elevates feedback from a post-mortem to a live metric. Each module generates performance data, risk signals, and usage context that feedback into the matrix. This feedback is then visualized on dashboards used by executive teams, compliance officers, and system designers alike. It has become a common language, a shared truth. Furthermore, in an age of complex, decentralized intelligence, that kind of shared situational awareness is not just helpful, it is essential.[12]

What the ECTM brings to modular intelligence is not constraint but clarity. It gives organizations the confidence to move fast because they have developed the ability to think critically. It aligns intelligence design with consequence design. It ensures that even as we decentralize cognition into modular systems, we centralize accountability where it belongs: in the hands of human stewards of the Hybrid Mind.

CHAPTER SUMMARY

Modular intelligence is not just a technical upgrade; it is a philosophical departure from the monolithic illusions of generality and scale. By coordinating specialized components, such as small language models, symbolic rule engines, and adaptive orchestrators, we build systems that are more resilient, transparent, and human-centered. These modular architectures thrive not in isolation but through thoughtful orchestration, where leadership shifts from micromanaging models to designing meaningfully distributed workflows.[2] We have seen how this design principle plays out across real-world use cases, including cognitive augmentation for patients with aphasia, adaptive accessibility pipelines for learners, and fairness-aware agents in multilingual interfaces.[3] Each case underscores a central truth: Modularity allows us to match intelligence to intention.[1]

At the strategic level, modularity is also a form of governance. The Experimentum Crucis Technology Matrix (ECTM) ensures that even agile, composable systems remain ethically grounded and organizationally aligned.[5] Through its scoring system, ECTM transforms innovation from a race into a rhythm, where experimentation is encouraged, but only if it stays accountable to human values and operational context.[6] In this modular landscape, feedback becomes not just a loop but a living signal. The system learns from itself. The orchestrator adapts. Furthermore, the human remains not just in the loop, but over it.[7]

As we transition to Chapter 6, we shift our lens inward. If modularity addresses how we build with intelligence, Synthetic Syndrome confronts how we live beside it. It explores the cultural and emotional undercurrents that arise when machines write, design, and speak with fluency, leaving humans to question their place in the loop. What happens when augmentation feels like displacement? What emerges when collaboration feels like competition? If Chapter 5 mapped the architecture of modularity, Chapter 6 delves into its psychological shadow, revealing the quiet dissonance that resides within the Hybrid Mind.

REFERENCES

1.   Gary Marcus and Ernest Davis, *Rebooting AI: Building Artificial Intelligence We Can Trust* (New York: Pantheon, 2019).
      Marcus and Davis argue for combining neural and symbolic systems – core to the modular intelligence model advocated here. Their critique of "black box" AI systems lends weight to the chapter's focus on orchestration and modular explainability.

2. Herbert A. Simon, *The Sciences of the Artificial*, 3rd ed. (Cambridge, MA: MIT Press, 1996).
   Simon's work on near-decomposability provides the intellectual foundation for modular systems as manageable, adaptive, and intelligible. His theory supports the rationale for modular coordination, orchestration, and traceability throughout the chapter.

3. Katherine Aafjes-van Doorn, Justin Bowers, and Andrew Przeworski, "Pragmatic Augmentation of Mental Health Therapy Using AI," *Journal of Technology in Behavioral Science*, 2022.
   This article presents applied examples of modular AI in therapeutic environments, such as layered support for individuals with language impairments. It substantiates the cognitive augmentation use case in Section 5.4 with real-world modular deployment.

4. Lin Zhang et al., "Composition of Experts: Redefining AI Model Architectures," *AI Systems Journal*, 2024.
   This paper provides key evidence that expert models working in coordinated modular networks often outperform monolithic architectures. It supports the foundational argument of Section 5.1 and the system coordination discussed in Sections 5.2 and 5.4.

5. Joseph X. Ng and Andres Fortino, Strategic Innovation Governance: Identifying Crucial Use Cases via Experimentum Crucis Technology Matrix (ECTM) (2025).
   This white paper outlines the Experimentum Crucis Technology Matrix (ECTM), which is used in Section 5.5 to structure modular evaluation across ethics, scalability, compliance, and market risk. It offers a scalable governance framework for AI innovation.

6. European Commission, *Ethics Guidelines for Trustworthy AI* (Brussels: European Commission, 2019).
   The EU guidelines provide ethical grounding that informs this chapter's discussion of fairness, traceability, and accessibility. They support the regulatory relevance of modular design patterns, especially in Sections 5.4 and 5.5.

7. Joseph X. Ng, *EXAI Manifesto: Ethical eXcellence in Artificial Intelligence (EXAI) Manifesto*, Humainority.org, 2025, https://Humainority.org/exai-manifesto/.
   The EXAI Manifesto introduces lifecycle accountability and co-agency as essential for ethical AI orchestration. It is a philosophical backbone for modular governance models throughout Sections 5.3 to 5.5.

8. Amar Gadre et al., "MRKL Systems: Modular Reasoning, Knowledge and Language," *Stanford HAI*, 2022.
   The MRKL system showcases how modular reasoning engines can coordinate symbolic and statistical components in practical AI applications. It informs the architectural core of Section 5.2 and supports hybrid integration across the chapter.

9. Judea Pearl and Dana Mackenzie, *The Book of Why: The New Science of Cause and Effect* (New York: Basic Books, 2018).
   Pearl's work on causal reasoning strengthens the chapter's advocacy for rule-based modules that produce interpretable, auditable logic. This directly informs traceability standards in Sections 5.2 and 5.5.

10. McKinsey Global Institute (Jacques Bughin, Jeongmin Seong, James Manyika, Michael Chui, and Ramesh Joshi), "Notes from the AI Frontier: Modeling the Impact of AI on the World Economy," 2019.
    This report explains the economic momentum behind rapid AI scaling, which modularity can accommodate more ethically and efficiently than monolithic deployment. It contextualizes the strategic appeal of modular systems in Sections 5.1 and 5.5.

11. Daniel Kahneman, *Thinking, Fast and Slow* (New York: Farrar, Straus and Giroux, 2011).
Kahneman's two-system theory is referenced in Section 5.3 as a useful analogy for orchestrating fast-acting and slow-deliberating modules. His influence also echoes in the structural logic of ECTM in Section 5.5.

12. Joseph X. Ng, *Mastering IT Innovation: 90-Day Proof of Value Blueprint* (New York: SAGE Ensights LLC, 2024), https://a.co/d/07ZO4zW.
Ng's blueprint offers practical implementation strategies for deploying AI modules in proof-of-value projects. It complements the governance approach in Section 5.5 and the orchestration lens in Section 5.3.

6 Human ETHOS

Restoring Moral Presence in the Age of AI

As AI systems grow more sophisticated, their presence in the workplace begins to shape not just tasks, but minds. What unfolds is a subtle shift, psychological and cultural, that culminates in a phenomenon known as Synthetic Syndrome. Rather than focusing on job loss or displacement, the chapter examines a deeper erosion, of meaning, confidence, and authorship, as automation begins to simulate human expression and decision-making with increasing persuasiveness.[1]

We begin by naming this shift, framing Synthetic Syndrome as a rational, emotional response to proximity with machine-generated work that feels interchangeable with our own.[2] From here, we define the Human ETHOS as a formal modality, our capacity to interpret, contextualize, and embed moral weight into decisions. This uniquely human lens cannot be compressed into code.[3] It must be preserved as a central pillar of hybrid intelligence, a concept that refers to the symbiotic relationship between human and machine intelligence in the context of AI systems.[4]

As outlined in Table 6.1, the framework operationalizes Empathy, Truth, Humility, Ownership, and Stewardship as the five ethical pillars that anchor hybrid decision-making. These are not lofty ideals but lived disciplines, each one offering a deliberate point of resistance against the frictionless acceleration of automation. ETHOS is what allows a clinician to pause when the model's output feels wrong, what moves a product designer to ask, *"Why does this work for some but not for others?,"* and what enables an educator to recognize that a flawless essay might still hide a disengaged student. In each case, Human ETHOS restores context, insists on care, and brings karmic awareness back into the loop.

Rather than resisting AI, Human ETHOS complements it, offering the moral latency required to interrupt speed with care and to temper optimization with meaning. Where machine intelligence narrows toward probability, ETHOS reopens the space of *possibility*. Where automation scales output, ETHOS restores *context*. In hybrid systems, it functions as the conscience of orchestration, a structured Human-In-The-Loop (HITL) presence that must not be bypassed or collapsed for the sake of efficiency. It is not a brake, but a moral circuit-breaker. As systems grow faster, more confident, and more generative, Human ETHOS does not compete; it stewards. Not by dictating outcomes, but by insisting we pause long enough to ask the right questions, carry the weight of consequence, and remain accountable for what we choose to build.

 DOI: 10.1201/9781003648338-8

TABLE 6.1
Human ETHOS – A Moral Framework for Hybrid Intelligence

Pillar	Definition	Why-Based Guiding Question
Empathy	The ethical perspective that centers human emotion, social context, and unseen impacts.	Why does this matter to those affected, and how might they feel – beyond the data?
Truth	A perspective of clarity and transparency that resists distortion, bias, or performative accuracy.	Why is this true, and what are the consequences if we act on illusion?
Humility	The recognition of cognitive and systemic limits – valuing doubt, reflection, and the space to be wrong.	Why might we not know enough – and how can we remain open to what we've missed?
Ownership	The karmic principle of accountability – accepting responsibility for downstream effects.	Why am I responsible for the outcomes this system creates, intended or not?
Stewardship	A long-view ethic of care – preserving human relevance through design for feedback, consequence, and care.	Why will this matter tomorrow, and are we prepared to respond and repair?

Human Trust Alignment in the Hybrid Age

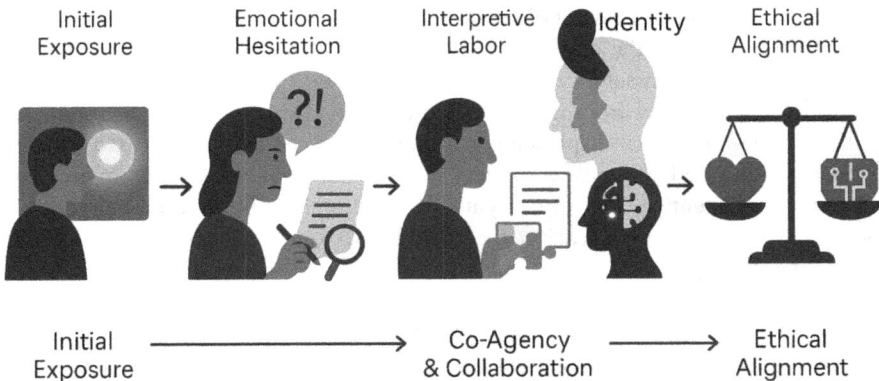

FIGURE 6.1   Human Trust Alignment in the Hybrid Age. Image generated using DALL·E 3 via OpenAI (ChatGPT v4.0) based on author-provided design prompts. Reviewed and edited for accuracy.

The chapter then shifts focus to the emotional labor involved in hybrid workflows (see Figure 6.1). As humans adjust, clarify, and reinterpret AI-generated outputs, they shoulder a growing, but often invisible, burden.[5] This interpretive work becomes crucial for maintaining trust, legibility, and connection in automated systems, particularly in sensitive domains such as healthcare (diagnosis and treatment planning),[6] education (student assessment and feedback), and management (decision-making and strategy formulation).[7]

We then examine how identity itself is affected. Imposter syndrome, in the AI era, is recast as identity drift, a condition where professionals no longer recognize themselves in their work.[8] When authorship is diluted and agency reduced to prompts and validation, the sense of being essential quietly erodes.[9]

Finally, the chapter introduces the concept of meaning preservation as a design principle. It offers tangible practices, such as hesitation layers (introducing pauses in AI processes to allow for human intervention),[10] co-authorship prompts (encouraging collaboration between humans and AI in creative tasks),[11] and emotional audits (regular assessments of the emotional impact of AI on human workers),[12] to ensure that human relevance is not only protected but also embedded into the logic of hybrid systems.[13] These practices frame the human not as an afterthought to automation but as the grounding presence that ensures what we build still reflects who we are.[14]

Together, these sections establish the emotional, ethical, and structural importance of the Human ETHOS, not just as a psychological safeguard but as a design foundation for the systems that follow. This prepares the reader to shift from internal orientation to external execution in the next chapter, where organizations move from seeking novelty to designing purpose-driven innovation through clearly framed, crucial use cases.[15]

6.1   NAMING THE HUMAN DISPLACEMENT

The first symptom of transformation is often a sense of unease. Not the kind that surfaces in headlines or shareholder meetings, but something quieter, an ambient doubt that creeps into the edges of competence. In this era of advanced AI, many professionals find themselves doing the work, meeting deadlines, and still feeling somewhat disconnected from the relevance of their efforts. This is not a matter of laziness or ignorance. It is displacement. Something else is doing what we used to do, faster, smoother, and, at times, eerily well. We are not obsolete, but we are no longer at the center of attention. The term "Synthetic Syndrome" captures this emerging psychological state, a creeping erosion of confidence not from failure but from proximity to machine fluency.[16]

Synthetic Syndrome does not begin with automation. It starts with admiration. The AI completes your draft. It suggests sharper phrasing. It finishes the code snippet. At first, the response is relief, even awe.

Nevertheless, over time, a strange internal shift occurs. If the system can create so effortlessly, what exactly am I contributing? Am I still necessary or merely supervisory? This is not paranoia. It is a rational reaction to intelligent systems that now participate in tasks we once believed were uniquely human. Writers, designers, educators, and marketers, those once defined by style, nuance, and tone, now watch as machines produce passable versions of their work. Not always excellent. However, it is increasingly good enough. That phrase, good enough, is where the fracture forms.

Recent studies suggest this effect is not hypothetical. In a 2024 feature, Psychology Today documented how professionals across various industries were reporting symptoms of imposter syndrome triggered not by human peers but by AI systems.[16] The syndrome is reframed: not "I am not good enough," but "Was I ever?" Forbes echoed this sentiment in a 2024 piece on AI-induced imposter feelings, emphasizing how AI's surface-level mastery often masks the absence of lived experience or purpose, but that distinction is not always emotionally legible.[12] Professionals know this intellectually. Nevertheless, emotionally, they are often left adrift.

This dissonance is especially pronounced when the AI output is not only technical but also expressive in nature. An LLM that generates poetry may not rival Shakespeare, but it does not need to. It only needs to be publishable. A generative design model may not invent an aesthetic movement, but it can convincingly remix one. In creative fields, where identity and output are deeply intertwined, this replication induces a subtle trauma. The artist begins to question the origin of their vision. The manager wonders if strategic insight is now just a matter of prompt engineering. In this landscape, the value of human contribution becomes unstable, not because it has disappeared but because it no longer feels distinct.

Within the Hybrid Mind framework, this moment is not dismissed, it is named. In the Preface and Synthetic Syndrome newsletter, Ng writes that this unease is not irrational but diagnostic. Feelings, he notes, are signals. Emotional disturbance in the workplace, particularly when humans feel "in the loop" but not "of the loop," is a sign that something fundamental is shifting.[15] The Hybrid Mind does not pathologize this reaction. It sees it as a necessary wake-up call to re-anchor human identity within hybrid systems. Because when machine outputs dominate attention, the invisible labor of being human, interpreting nuance, holding uncertainty, weighing ethics, becomes dangerously undervalued.[8]

This displacement is not confined to creatives. In healthcare, clinicians using diagnostic tools powered by deep learning often feel reduced to validators, rubber-stamping AI-recommended care paths.[1] In education, instructors are increasingly relying on grading assistants and adaptive feedback tools, which perform with remarkable fluency but often lack pedagogical expertise. In finance, anomaly detection systems now flag risks faster than teams can comprehend their cause.[4] Across each domain, the professional remains "in charge" yet often feels like a spectator to the machine's confidence. The danger here is not job loss. It is the erosion of meaning.

The displacement we are witnessing is not a glitch; it is a feature of a world moving too quickly to reorient. The issue is not merely technical capability but the emotional and existential lag that follows rapid automation. The Financial Times reports that workers interacting with high-autonomy AI systems experience higher levels of mental fatigue and lower satisfaction, not because they dislike the tools but because they lose their sense of authorship.[17] The danger, then, is not synthetic intelligence, it is synthetic detachment. When the machine gets smarter, we do not necessarily get dumber, but we do risk feeling smaller.

To move forward, we must first call it what it is: a quiet crisis of meaning. Not loud enough to trigger lawsuits. Not stark sufficient to halt progress. However, it is potent enough to alter the texture of work, the rhythm of identity, and the core of what it feels like to contribute. Synthetic Syndrome is not a fear of replacement. It is a longing for recognition in a system that increasingly mirrors our minds but cannot reflect our meaning. It is the early warning sign that in designing for speed, we may be sidelining selfhood.

6.2   HUMAN ETHOS AS THE LAST MODALITY OF MEANING

At the heart of every meaningful decision is not just a calculation but moral questions. This is the lens we call Human ETHOS: a question-based ethical compass grounded in empathy, truth, humility, ownership, and stewardship. These are not abstract ideals but lived disciplines, modes of orientation that resist reduction, protect relevance, and hold the loop open for human reflection in automated systems. It is not made of data but of experience. As artificial intelligence becomes more fluent, producing essays, art, and analysis with ease, the need to define what remains uniquely human becomes urgent. At this moment, the Human ETHOS emerges not as a vague ideal but as a formal modality, a distinct, non-replicable capacity to embed meaning, emotion, and moral orientation into context. Unlike AI, which is trained on patterns and probabilities, the human mind constructs narratives shaped by memory, emotion, and ethical weight. This chapter proposes that Human ETHOS is not merely a cognitive asset, it is the last irreducible modality of intelligence.[7]

Machines can simulate relevance, but they do not feel it. They can generate context, but they do not inhabit it. The distinction is not trivial. Human ETHOS is constructed through layers of orientation, our values, biases, lived experiences, and the cultural scaffolding that guides how we see the world. As outlined in Chapter 2's exploration of the OODA-F loop, orientation is the most subjective and yet most essential component of decision-making. Without it, decisions may be fast and logical, but they are rarely wise. When AI systems bypass or flatten the orientation layer, what is lost is not just insight but humanity. A diagnosis loses compassion. A recommendation loses ethics. A conclusion loses consequence.[6]

Fei-Fei Li, a leading voice in human-centered AI, has noted that our priority must be ensuring AI "enhances human capabilities rather than replacing them."[13] This requires more than technical design. It demands a recognition of what cannot, and should not, be optimized. Human ETHOS thrives on ambiguity. It tolerates contradiction. It chooses presence over precision when the situation calls for it. No model can learn this through data alone because these qualities are not statistical, they are existential.[18]

In many high-functioning workplaces, the erosion of this modality is already visible. Consider how generative AI tools, such as ChatGPT or Midjourney, can produce corporate communications, branding collateral, or strategic plans in a matter of moments. The results may be accurate and well-formed but often lack soul. They mimic tone but miss tension. They flatten dissent. They exclude the friction that gives human expression its edge. Professionals who rely too heavily on these systems report not just creative fatigue but narrative disorientation, a sense that their voices are becoming interchangeable. According to Time magazine's 2025 feature on relational automation, even therapists and caregivers working with emotionally responsive bots have begun to question the authenticity of their reactions as machines increasingly mirror empathy without experiencing it.[17]

This is not a call to fear AI. It is a call to re-center humanity within systems we co-create. Ng's EXAI Manifesto emphasizes this orientation as both a right and a responsibility.[15] If the future is hybrid, then the role of humans is not just to supervise machines, it is to provide the emotional grounding they inherently lack. That includes knowing when the context should not be compressed when speed is not wisdom, and when silence is more truthful than certainty. This is where Human ETHOS becomes not just a defense against automation fatigue but a design principle in its own right.[5]

To preserve this modality, we must acknowledge its structural importance. Emotional intelligence, intuition, narrative sensemaking, and ethical orientation are not "soft" skills, they are synthesizing skills. They do what data alone cannot: they create meaning across time. As the Financial Times has reported, employees embedded in highly automated systems often feel disconnected from outcomes despite hitting performance metrics.[17] The output is technically sound but personally hollow. This detachment signals a loss of authorship, not of tasks, but of identity. If we cannot see ourselves in the process, we will soon lose interest in its result.

What remains irreducibly human is not our capacity to produce outputs but to care about them. The machine may write the report, but only the human can decide what it means, and what it costs. In a world of accelerating capabilities, the Hybrid Mind must not just adapt faster. It must remember slower. It must hold the line of perspective not as nostalgia but as necessity. Because only from that vantage point can the work we do still belong to us.

6.3   EMOTIONAL LABOR AND QUIET WORK OF INTERPRETATION

There is work that leaves no timestamp. No tracked revision. No log in the system. Nevertheless, it is everywhere, especially in hybrid environments where humans interact with AI-generated content. It is the work of interpreting tone, smoothing ambiguity, and translating machine logic into something relational. This is emotional labor, and in the age of AI, it is multiplying. Not because machines are emotional but because they are not. The more fluent AI becomes at outputting prose, images, or decisions, the more human labor is required to ensure those outputs are placed in context, with care, with calibration, and with consequences considered. This is the quiet work of interpretation.

Sociologist Arlie Hochschild first coined the term "emotional labor" to describe how workers, particularly in service roles, manage their emotions to fulfill the requirements of their jobs.[8] What she observed in flight attendants and call center staff now applies to knowledge work across the board. Emotional labor today often involves adjusting or validating something an AI-generated text has written. For instance, when a generative tool drafts a follow-up email that's too direct, a human rewrites it for diplomacy. When a résumé screening system scores a candidate low based on language that misses nuance, the hiring manager manually intervenes. When a recommendation engine suggests a product, the marketer adjusts the framing to avoid tone-deafness. These are not just edits. They are interpretive acts that restore humanity to otherwise mechanical exchanges.

The scale of this labor is growing, but its visibility is not. In many organizations, employees act as unacknowledged buffers between AI systems and end-users. They catch the errors, soften the edges, and make sure the machine does not sound like a machine. In doing so, they preserve trust, but often at the cost of personal cognitive load. A 2025 article in The Financial Times highlights this emerging toll: workers in high-automation roles report feeling mentally drained, not because the work is hard, but because the interpretive burden is heavy and unrewarded.[17] The labor is not about doing more. It is about meaning more, which is harder to measure and easier to overlook.

Emotional labor also takes place in teams managing AI implementation. When engineers deploy a chatbot that misinterprets sentiment, it is the frontline support agent who must apologize, clarify, and rebuild rapport with the customer. When a model recommends layoffs based on performance metrics, it is the manager who must deliver the message, absorb the anger, and carry the moral weight. AI may optimize efficiency, but it outsources the emotional consequences to humans. This reconfiguration of responsibility is subtle but significant. The Hybrid Mind must recognize this redistribution and design for it, not just around it.

Williams, writing at the time, noted that when AI enters caregiving professions, whether in eldercare robots or emotionally responsive customer service avatars, it can mimic empathy well enough to deceive users. However, the humans who maintain these systems often experience emotional whiplash.[17] They must toggle between machine logic and human emotion, interpreting one for the sake of the other. Over time, this kind of interpretation becomes a silent toll on self-worth and relational clarity. Emotional labor, then, is not just about softening the machine. It is about preserving the human.

What makes this labor especially precarious is that it is rarely acknowledged as a skill. It lives in intuition, gut checks, and interpersonal tact. Nevertheless, it is precisely this interpretive ability that keeps hybrid systems trustworthy. Ng frames this in the EXAI Manifesto as the essential role of the human-in-the-loop, not just to approve outcomes but to inject context where none exists.[15] Emotional labor becomes infrastructure. It is not decorative or optional. It is what holds the system accountable to the lives it affects.

In technical terms, one might refer to this as "meaning alignment." However, in lived terms, it is what people do when they read between the lines. When they feel something is off. When they take a breath before clicking send, AI may propose. But humans compose. Moreover, composing is not just an act of expression. It is an act of empathy. Of ethics. Of perspective.

The Hybrid Mind must, therefore, protect emotional labor from becoming emotional leakage. That means not only acknowledging its existence but embedding it into system design. That includes feedback loops that surface human edits as signals, not errors. It includes rituals in a workflow that make space for interpretive reflection, not just correction. Furthermore, it includes redefining productivity, not as speed, but as significance.

Because when machines generate language, it is humans who translate it into relationships. That is not a small task. It is the task that makes all the others matter.

6.4   IDENTITY DRIFT IN THE MIRROR OF MACHINES

Imposter syndrome is not new. It has long haunted high achievers who doubt their accomplishments, attributing success to luck or external validation rather than their internal ability. However, in the AI age, the syndrome mutates. It shifts from insecurity about one's competence to confusion about one's role. When a system can produce in seconds what once took hours of care, what happens to the meaning of mastery? When the machine outputs work that appears polished, efficient, and confident, many professionals look at their efforts, and then at the AIs, and begin to wonder not "Am I good enough?" but "Am I even relevant?"

This phenomenon is what we refer to as identity drift. It is not a sharp break but a slow dissolve. The edge between human and machine contribution becomes so blurred that professionals start to dissociate from their work. Designers who once labored over layouts now click through autosuggestions, writers who once crafted their voice now scan through generated options. Strategists who once charted direction now rely on predictive dashboards. The machine performs, and the human begins to feel more like a reviewer than a creator. The result is a hollowing out, not of skill, but of selfhood.

This is not just a theoretical risk. In a 2024 Psychology Today feature, clinical psychologists reported a measurable uptick in clients expressing a new kind of creative fatigue, one rooted not in overwork but in comparison to AI-generated outputs.[16] These individuals were not burned out in the traditional sense. They were disoriented. They felt their voice was being replicated without being needed. Their ideas were "close enough" to what the machine could suggest. The symptom was not failure but interchangeability. They were not losing their jobs. They were losing their distinctiveness.

The literature supports this shift. A 2025 Forbes analysis reframes imposter syndrome as increasingly situational and externally induced, triggered by proximity to the system rather than personal doubt.[12] The AI does not try to outshine you. However, by doing its job well enough, it makes you question whether your contribution is essential or merely habitual. For many, this is not a crisis of competence. It is a quiet grief. There is a sense that their work still gets done, but their presence in that work has faded.

Ng, writing in the EXAI Manifesto, argues that the right to authorship is a fundamental human need. Not authorship in the legal sense but in the narrative one, the sense that one's choices shape the trajectory of outcomes.[15] This is where identity forms: in causality, not just capability. The AI may complete your sentence, but it does not know what it costs you to say it. It cannot locate you in the tension, the uncertainty, or the moral weight of your decision. Only you can. That is the function of the Human ETHOS. Furthermore, when systems operate without space for that function, identity begins to drift.

To preserve authorship, professionals are beginning to develop rituals of resistance. Some deliberately slow down workflows, refusing to accept the first AI draft. Others inject ambiguity into prompts to reclaim interpretive space. Still, others mute autocomplete features or insist on hand-crafting parts of presentations. These are not technophobic gestures. They are identity restorations. They are ways of saying, "I was here. I shaped this." Not because the machine could not, but because the meaning would not be the same without human touch.

The Hybrid Mind does not dismiss imposter feelings. It interrogates them. It understands that doubt is not always irrational, it is often a signal that something is amiss. A system warning that an internal component is misaligned with the interface's representation. Identity drift, in this sense, is a diagnostic, not of weakness, but of compression. When authorship is flattened, we feel it. When an agency is automated away, we miss it. This is not nostalgia. It is memory, the memory of a time when we were not just involved in the process but responsible for it.

To lead in an age of intelligent systems, we must design for authorship, not just accuracy. Because what humans crave is not just being right but being real.

6.5   DESIGNING FOR MEANING PRESERVATION

In the face of synthetic fluency, speed, and scale, it is tempting to accept AI's outputs as sufficient. The logic is persuasive: if the model generates the draft, completes the design, or proposes the strategy faster and with acceptable quality, then why hesitate? However, what is acceptable is not always what is meaningful. The final evolution of Synthetic Syndrome is not exhaustion, it is numbness, a quiet detachment from the very work we once found fulfilling. Meaning is not something AI can preserve for us. It is something we must actively protect. That protection begins with design.

Meaning preservation is the conscious practice of embedding human interpretation, authorship, and ethical orientation into workflows that increasingly rely on automation. It is not a resistance to AI but a recalibration of pace, responsibility, and presence. At its core, it means slowing down where it matters, not across the board, but at strategic inflection points where judgment, emotion, or stakes converge. For example, a healthcare AI may flag a risk pattern in a diagnostic scan, but the conversation with the patient – the framing, the delivery, the timing – is where meaning is shaped. That cannot be automated.[1]

In EXAI-aligned systems, this design principle translates into three practical interventions. The first is the use of hesitation layers, which involve intentionally embedding pauses into automated decision flows, allowing human reviewers to reflect before finalization. These can take the form of confirmation prompts, ethical review checklists, or real-time transparency flags. Rather than being seen as friction, they serve as narrative interludes, moments to ask not just what is happening but why it matters.[10]

The second is co-authorship prompts. These are system nudges that invite human input not as a correction but as a contribution. Instead of suggesting a rewrite, the AI might ask, "Would you like to emphasize empathy here?" or "Does this language reflect your values?" These are not grammar fixes. They are meaning invitations. They remind the human user that authorship is not just allowed, it is expected. In systems like Humainority or collaborative creative tools, these features become scaffolding for human voice rather than containers for machine prediction.[5]

The third is emotional audits. This practice acknowledges that humans working alongside AI are not just interacting with data, they are carrying emotional weight. These audits do not assess emotional states algorithmically. They offer reflective space. Short journaling boxes, pulse-check prompts, or team debrief moments embedded into the workflow give people a moment to re-enter themselves. As Hochschild described decades ago, emotional labor is real labor.[8] Furthermore, in hybrid systems, it becomes more critical, not less.

This means preservation also has a spatial dimension. Just as we design physical workplaces to reflect culture, we must design digital systems to reflect presence. In The Hybrid Mind, Ng emphasizes that hybrid leadership is not about resisting auto-mation but about maintaining authorship over systems that increasingly suggest the next move.[15] That authorship cannot survive if every move is frictionless. Systems without space for reflection produce output, but not orientation. Moreover, without orientation, relevance decays.[6]

In a recent *Time* article on the impact of AI on caregiving professions, technologists and clinicians alike raised concerns about the loss of meaning in human-machine interactions.[17] When emotional nuance is simulated too well, there is a risk that people stop seeking genuine empathy. When synthetic care becomes indistinguish-able from human concern, the relational spark begins to dim. Meaning preservation, in this context, is not just about the worker, it is also about the recipient. Systems must be designed not just for utility but for dignity.

The Hybrid Mind is not a nostalgia project. It is a future project grounded in per-spective. It recognizes that what makes human work worth doing is not how fast it happens but how much of ourselves we recognize in the result. Meaning cannot be retrofitted. It must be designed from the beginning.

To design for meaning is to protect the signal of humanity in a world of acceler-ating patterns. It is to embed authorship where automation tempts us to coast. It is to insist, gently but firmly, that the work we do still reflects who we are, and who we aspire to be.

6.6   HUMAN ETHOS FOR EXECUTIVE LEADERS

In the preceding sections, we treated Human ETHOS as a moral posture, a personal practice of care, judgment, and accountability within systems increasingly driven by autonomous computation. But in real-world environments, individual conscience doesn't scale. Ethical reflection at the edge must be mirrored by ethical clarity at the center. That center is executive leadership.

In hybrid systems, where decisions are shared between humans and machines, the role of the executive is no longer limited to setting strategy and approving roadmaps. Leadership becomes an act of design itself. Every policy, every product decision, every algorithm authorized by a board or C-suite has the potential to generate impact far beyond its immediate intention. In this context, Human ETHOS must be translated from a humanist framework into what we might call a moral operating system for lead-ership, one deeply conscious of karma, not as mysticism, but as systemic feedback.

TABLE 6.2
Human ETHOS for Executive Leaders

ETHOS Principle	Executive Interpretation	Strategic Function	Moral Perspective
Empathy	Stakeholder-centered insight and ethical foresight	Identifies unseen risk; strengthens product-market trust	Expand perspective to include the ignored; design for emotional karma
Truth	Transparency in assumptions, signals, and outputs	Safeguards brand integrity; enhances regulatory confidence	Speak with clarity; karma begins where distortion hides
Humility	Ethical restraint and recognition of cognitive/systemic limits	Supports checks, pauses, and epistemic transparency	Leave space for doubt; when certainty dominates, humility safeguards direction
Ownership	Executive accountability for downstream system effects	Clarifies responsibility; enables traceability and trust	Accept karmic consequence beyond legal scope; claim the ripple
Stewardship	Long-view thinking with structural feedback loops	Future-proofs design; aligns with cultural and temporal shifts	Perspective over time; karma returns through systems we don't yet see

Building on this framework, Table 6.2, translates the five pillars into executive-level practices, offering leaders a decision discipline that integrates moral presence with strategic oversight. Each of the five pillars of ETHOS – Empathy, Truth, Humility, Ownership, and Stewardship – can be reframed not just as values but as functional principles of executive decision-making. Empathy, for instance, becomes more than compassion; it becomes a leadership competency for surfacing unseen risk. An empathetic executive doesn't merely sympathize with users; they require their teams to design with marginalized stakeholders, to ask who is left out of the data, who is least protected by the system's default behaviors. It is a posture of expanded human perspective, an effort to see beyond dominant use cases and toward emotional and cultural edge cases. It is a posture of proactive proximity, ethical risk detection as an act of business intelligence, ensuring you are always prepared for the ethical challenges that lie ahead.[12]

Truth, in an executive context, is not reducible to factual accuracy. It is the integrity of signals, the clarity of assumptions, and the willingness to speak plainly when models become more persuasive than they are right. A CEO who misrepresents the capabilities of an AI product not only misleads the market but licenses a culture of distortion throughout the organization. In contrast, an executive who demands epistemic transparency, who insists that teams disclose uncertainty, edge cases, and model limitations, sends a different message: trust is built not by perfection, but by truthfulness with consequence.[6]

Humility, third in the ETHOS loop, carries strategic weight in environments increasingly driven by speed and statistical confidence. Executives are often urged to "trust the system" or "let the data lead," but humility resists blind deference to automation. It acknowledges that models, no matter how optimized, are fallible – and that success metrics can obscure harm. When an algorithm maximizes engagement at the cost of addiction, or when a hiring tool amplifies historical bias, humility is the quality that pauses, questions, and redirects. Leadership here means holding the authority to say, "The system is performing, but it may not be right," and then to act accordingly.[7]

Ownership is where many ethical failures collapse. Diffused systems tend to produce diffused responsibility. Suppose no one claims the impact, no one answers for it. Human ETHOS at the executive level reasserts that leadership is, fundamentally, a willingness to be accountable for systems one may not have personally coded or directly deployed. It is not a matter of legal liability, but moral authorship. This becomes critical when things go wrong, as they inevitably will. An executive who views ownership as reputation management will act defensively. But one who sees ownership as a long horizon of consequence will build systems with traceability, escalation paths, and post-deployment ethics as an operational norm.[4]

Stewardship, finally, is the frame that expands executive decision-making beyond the quarterly cycle. In its deepest form, it asks leaders to design with futures in mind that they may not personally witness. It calls on them to treat systems not as products, but as infrastructures, sociotechnical environments that people must live in, depend on, and be shaped by. Embedded within this is Feedback, not just as customer reviews or market signals, but as a structural ethic. Feedback is how leadership listens systemically. It is how organizations detect slow harms, edge-case failures, and emergent behavior that was neither modeled nor intended. Designing for feedback is an act of humble perspective. It reflects a willingness to remain in dialogue with karma, to be shaped by what your systems teach you, not just what you teach them.

When these pillars are taken together – Empathy for the unseen, Truth in signal, Judgment as interruption, Ownership as moral authorship, and Stewardship with feedback – Human ETHOS becomes more than an ethical frame. It becomes a strategic tool for governing complexity. For executives, it offers a robust structure of thought to navigate decisions that involve both technical ambiguity and moral weight. Not every decision will be clean. But every decision can be deliberate, and with Human ETHOS as your guide, you are equipped to make those deliberations.

This, then, is Human ETHOS at the executive level: not a virtue signal, not a compliance shield, but a compass. In systems of growing intelligence, leadership becomes ethical design by another name. And the leader who does not ask the moral question, why this, for whom, with what consequence, is no longer leading, but merely approving. Perspective is the capacity to see the others affected. Karma is the certainty that consequence will circle back. ETHOS is the reliable compass that guides you to lead with presence, with clarity, and with care, even in the most complex of decisions.

CHAPTER SUMMARY

In an age defined by scale and simulation, the human thread is more easily lost than ever. Synthetic Syndrome, emotional fatigue, and identity drift are not fringe experiences – they are early signals of deeper systemic misalignments.[16] This chapter repositions those emotional undercurrents not as obstacles to innovation but as guideposts for restoring authorship, empathy, and significance in hybrid systems. At the center of this restoration is the Human ETHOS: a meaning-making lens through which data becomes relevant, decisions regain moral weight, and consequences are acknowledged through a karmic loop.[7]

What we preserve is not just our ability to contribute, but our capacity to care about what we contribute. The Hybrid Mind, as a design and leadership posture, steadfastly resists the temptation to move fast and depersonalize. Instead, it slows where it must,[10] intervenes where it counts,[15] and holds open the space where human presence makes the difference between mere output and actual outcome.[6]

Yet insight without structure rarely scales. In the final section of this chapter, we reframed Human ETHOS not just as an individual compass but as an executive operating system, a perspective-aligned, karma-conscious framework for governing intelligent systems. At scale, empathy becomes risk detection. Truthfulness becomes a trust asset. Judgment becomes a moral override. Ownership becomes traceable authorship. And stewardship becomes system feedback tuned for generational resilience.

If Human ETHOS is the compass, then our maps must evolve. Chapter 7 transitions from emotion to execution, from preserving meaning to building systems that reflect it. This shift is not just a change in direction, but a call to action, a focus on the organizational imperative: not to chase killer apps, but to define crucial use cases. What matters is not just what we can build, but what we should.[19] The path forward demands discernment. Let us draw it together.

REFERENCES

1. Aafjes-van Doorn, Katja, Jesse Bowers, and Andrew Przeworski. "Pragmatic Augmentation of Mental Health Therapy Using AI." *Journal of Technology in Behavioral Science*, 2022.
    This study illustrates how AI is increasingly present in therapeutic settings, revealing new forms of emotional labor required by human practitioners. It supports the chapter's argument that interpretive and affective labor is not eliminated – but intensified – in hybrid roles.
2. Bousquette, Ian. "Johnson & Johnson Pivots Its AI Strategy." *The Wall Street Journal*, April 18, 2025.
    This case study offers real-world insight into how increased AI adoption impacts employee identity and confidence. It reflects the corporate-level implications of Synthetic Syndrome and identity drift.
3. Chomsky, Noam. *Aspects of the Theory of Syntax*. MIT Press, 1965.
    Chomsky's theory distinguishing deep structure from surface structure provides theoretical grounding for why AI outputs may sound human but lack contextual

understanding. This distinction bolsters the chapter's definition of the Human ETHOS as irreducible.

4. Deloitte. State of AI in the Enterprise: 5th Edition. Deloitte Insights, 2023.
This industry report documents the growing emotional and cognitive burden on workers tasked with supervising intelligent systems. It provides empirical support for the chapter's claim that automation often creates invisible oversight labor.
5. Fei-Fei Li. "Human-Centered AI: Building Trust with Empathy and Ethics." *Stanford HAI Lecture*, 2021.
Fei-Fei Li emphasizes that systems designed without empathy or ethical awareness risk losing human trust. Her vision informs the chapter's argument for co-authorship prompts and hesitation layers.
6. Floridi, Luciano, and Josh Cowls. "A Unified Framework of Five Principles for AI in Society." *Nature Machine Intelligence* 1, no. 1 (2019): 65–67.
Floridi and Cowls propose that explicability and human agency must be built into AI systems. These principles directly support the ethical framing of the Human ETHOS as a system design requirement.
7. Gardner, Howard. *Five Minds for the Future*. Harvard Business School Press, 2007.
Gardner's concepts of the respectful, ethical, and creating minds provide conceptual scaffolding for the Hybrid Mind model. They validate the chapter's assertion that meaning-making and moral discernment are essential forms of intelligence.
8. Hochschild, Arlie Russell. *The Managed Heart: Commercialization of Human Feeling*. University of California Press, 1983.
Hochschild's foundational work on emotional labor is extended in this chapter to AI-augmented professions. Her analysis supports the call for emotional audit practices in human-machine collaboration.
9. Horvitz, Eric. Quoted in *The Hybrid Mind: The Human-AI Convergence*, 2025.
Horvitz's observations on the structural risks of automation help frame Synthetic Syndrome not as human fragility, but as a systemic warning sign. His perspective validates the urgency of preserving human emotional and contextual framing.
10. Kahneman, Daniel. *Thinking, Fast and Slow*. Farrar, Straus and Giroux, 2011.
Kahneman's dual-system model of decision-making supports the chapter's call for hesitation layers and slower reflection in high-speed AI environments. His work reinforces the idea that interpretive friction is not inefficiency – it is insight.
11. Kelly, Kevin. "Humans Are the Sexual Organs of Technology." *Wired*, 2005.
Kelly's provocative metaphor positions humans as creators of technologies that evolve beyond them, illuminating the disorientation felt when machines begin mimicking or replacing human expression.
12. Lambert, Anna. "Overcoming AI-Induced Imposter Syndrome." *Forbes*, January 10, 2025. www.forbes.com/sites/annalambert/2025/01/10/overcoming-ai-induced-imposter-syndrome/
Lambert captures the growing phenomenon of AI-induced imposter syndrome in knowledge workers. Her reporting aligns with the chapter's reframing of this syndrome as structurally induced rather than individually pathological.
13. Li, Fei-Fei. "Human-Centered AI: Building Trust with Empathy and Ethics." *Stanford HAI Lecture*, 2021.
Fei-Fei Li's insights are referenced again here to emphasize that empathy and ethical alignment are central to meaningful human-AI interactions. She supports the design philosophy underlying Human ETHOS preservation.

14. McGilchrist, Iain. *The Master and His Emissary: The Divided Brain and the Making of the Western World.* Yale University Press, 2010.
    McGilchrist's hemispheric brain model helps frame the internal tension many humans feel in hybrid work, particularly between analytic acceleration and holistic coherence. His work deepens the neurological metaphor for identity drift.

15. Ng, Joseph X. *EXAI Manifesto: Ethical eXcellence in Artificial Intelligence (EXAI) Manifesto.* Humainority.org, 2025. https://Humainority.org/exai-manifesto/
    This manifesto outlines practical safeguards – such as hesitation layers, emotional audit trails, and authorship rights – used throughout Chapter 6. It supplies the ethical and architectural backbone for designing human-centered systems.

16. Nosta, John. "AI and the New Impostor Syndrome." Psychology Today, The Digital Self (blog), updated March 13, 2025. Reviewed by Lybi Ma. www.psychologytoday.com/us/blog/the-digital-self/202503/ai-and-the-new-impostor-syndrome
    Nosta discusses how generative AI systems are undermining creative professionals' sense of contribution and authorship. Her reporting grounds the chapter's concept of "identity drift" in lived psychological experience.

17. Williams, T. "When AI Automates Relationships." *Time*, March 2, 2025. https://time.com/7010288/when-ai-automates-relationships-essay/
    Williams captures the relational fatigue experienced by professionals in emotionally mediated roles. Her insights lend credibility to the chapter's argument that machines simulating empathy can displace the human need to be empathetic.

18. Spiegelhalter, David. *The Art of Statistics.* Pelican, 2019.
    Spiegelhalter emphasizes the narrative and ethical responsibilities behind data interpretation. His work supports the idea that numbers and model outputs do not carry meaning on their own – only humans can confer that.

19. Vallor, Shannon. *Technology and the Virtues: A Philosophical Guide to a Future Worth Wanting.* Oxford University Press, 2016.
    Vallor's techno-ethical philosophy frames moral character as central to good technology design. Her vision informs the chapter's argument that AI systems must be built to support – not replace – human discernment.

Part III

Governance and Risk
in Exponential Systems

MOVING FROM EXPERIMENTATION TO DISCERNMENT

In the face of exponential technological acceleration, experimentation is no longer a luxury, it is a necessity. However, unchecked experimentation without discernment breeds chaos, redundancy, and risk. Part III of this work argues that responsible innovation in the AI era requires a shift from enthusiasm to evaluation, from pilots to principles, and from invention to intentional oversight. Governance becomes the connective tissue between potential and practice, linking technological ambition with human accountability.

The modern enterprise suffers from what has been called the innovation bottleneck. Organizations rush to trial emerging technologies but fail to scale the ones that matter. As companies adopt hundreds of AI use cases, most of which remain trapped in sandbox mode, only a fraction ever deliver meaningful impact. This pattern of diffusion without discipline is driven by a fundamental misconception: that more experimentation automatically leads to progress. It leads to saturation, confusion, and, in many cases, erosion of trust. To counter this, we must turn our attention to what Joseph Ng and Andres Fortino define as "crucial use cases," those rare, high-leverage opportunities where AI is not just functional but foundational to solving a real problem.[1]

The concept of the experimentum crucis, drawn from the history of science, offers a helpful metaphor. It refers to the critical experiment that decisively proves one hypothesis over another. In the business context, it means surfacing the killer use, the compelling, often latent problem that, when addressed by the right technology, makes adoption not only obvious but inevitable. This is not about chasing hype or trying every new tool. It is about discovering where value is waiting to be unlocked. When framed this way, governance becomes an instrument of clarity, not a constraint. It helps leaders distinguish between noise and signal, between interesting ideas and necessary transformations.[2]

The Experimentum Crucis Technology Matrix (ECTM) provides the structure to enable that discernment. It is not simply a checklist but a multidimensional evaluation tool that guides decision-makers in understanding whether a use case is ready, ethical, scalable, and safe. The matrix encompasses dimensions such as technology readiness, business impact, compliance, cybersecurity, workforce implications, ESG alignment, and market risk. Together, these dimensions form a decision model rooted in both strategic and moral logic. For example, a generative AI model proposed for legal document summarization might score high on productivity and readiness but low on interpretability and risk mitigation. In such cases, ECTM helps organizations pivot or redesign before they are exposed to real-world consequences.[2]

Governance is not a brake; it is an accelerant for trust. This becomes especially evident when we consider the role of audits. Traditionally, audits are backward-looking reviews of compliance. However, in AI governance, audit must be reimagined as a real-time design mechanism. This is the premise behind the Hybrid Audit Model (HAM), a lightweight oversight framework proposed in conjunction with the ECTM. HAM maps audit logic onto the OODA loop – Observe, Orient, Decide, Act – embedding reflection directly into operational systems. In this reframing, audit is not something that happens after deployment; it is baked into the loop, allowing for dynamic adjustments, ethical recalibration, and the early detection of failure modes.[2]

The risks are not merely technical. There are also social and psychological risks. Synthetic content generation blurs the line between authentic and artificial. As more content is machine-generated, the informational ecosystem is saturated with simulations. This collapse of meaning, a condition explored in Ng's theory of Synthetic Syndrome, is not just an epistemological concern but a governance failure. If humans cannot distinguish between what is real and what is mimicked, we lose the capacity to act with informed agency. The governance of exponential systems must, therefore, include cultural and cognitive safeguards. This means watermarking AI outputs, designing friction into automation, and preserving human interpretability even when speed tempts us to bypass it.

Governance is also a leadership function. In a hybrid world, leaders are not just decision-makers, they are ethical architects. They must define boundaries, set escalation paths, and create incentives for responsible innovation. Without this, organizations risk tech debt not just in infrastructure but in trust. Scaling AI without such safeguards is akin to building skyscrapers without blueprints. The ECTM and HAM frameworks provide those blueprints, but they must be activated through disciplined practice. This includes scenario testing, continuous scoring, and feedback loops that surface both risks and reflections.

As companies scale AI, they will need to confront the trade-offs between centralization and decentralization, autonomy and oversight, speed and safety. Governance is what holds these tensions in productive balance. It is how discernment replaces experimentation as the hallmark of innovation maturity. The Hybrid Mind does not merely build intelligent systems; it builds systems that remember what humans value and are willing to protect.

REFERENCES

1. Joseph X. Ng and Andres Fortino, "Strategic Innovation Governance: Identifying Crucial Use Cases via Experimentum Crucis Technology Matrix (ECTM)" (2025).
   This paper introduces the ECTM framework, grounding it in the concept of "crucial use cases" and providing a structured governance method to scale innovations responsibly. It serves as the conceptual and operational foundation for Part III's exploration of innovation governance.
2. Cătălin Dumitru, "Crucial Instances and Crucial Experiments in Bacon, Boyle, and Hooke," *Society and Politics* 7, no. 1 (2013): 45–53.
   Dumitru's work contextualizes the philosophical origins of *experimenta crucis*, offering a historical lens for understanding the ECTM's decision-theoretic orientation in enterprise innovation.

7 From Killer Apps to Crucial Use Cases

In a culture hooked on the thrill of the next "killer app," it is easy to mistake motion for meaning. But what if innovation is not about chasing the next big thing, but about asking better questions? Here, the spotlight shifts inward, toward a more profound reckoning with how we define progress and what it takes to build something that endures. In the early days of personal computing and the internet, killer applications like VisiCalc and Netscape played a pivotal role in platform adoption.[2] However, in the era of artificial intelligence, this mindset, fueled by hype cycles and demo-driven development, has become a liability.[1] Instead of focusing on singular breakthroughs, the Hybrid Mind prioritizes "crucial use cases": technology applications grounded in real problems, aligned with long-term organizational values, and governed by ethical, operational, and social clarity.[3] This shift in focus presents a promising future for AI innovation, one that is grounded in real-world problems and guided by ethical principles.[4]

We begin by deconstructing the myth of the killer app, exploring how it distorts investment decisions and undermines responsible scaling. From there, the chapter introduces use cases as governance anchors, structured, intentional scenarios that guide innovation with accountability and purpose.[5] Drawing on the Experimentum Crucis Technology Matrix (ECTM), the discussion illustrates how reframing innovation through use cases can reorient teams toward creating longitudinal value rather than pursuing short-term novelty.[4]

The chapter then outlines the full lifecycle of a use case, from framing and scoping to piloting, scaling, and sustaining, illustrating how innovation must be nurtured, stress-tested, and continually recalibrated.[6] Real-world examples, such as BNY Mellon's predictive ML models and Humainority stealth résumé optimizer, a tool designed to optimize job applications by removing potential biases, are used to demonstrate this lifecycle in practice.[7]

We close with a call to action for the Hybrid Innovator. Instead of seeking viral success, the Hybrid Innovator leads with thoughtful questions: Who is impacted? What assumptions are being codified? What parts of this system must remain human? These questions serve as the scaffolding for ethical and practical innovation in complex systems.[8] This shift in mindset from seeking viral success to asking thoughtful questions can inspire a change in your approach to innovation.

DOI: 10.1201/9781003648338-10

Chapter 7 marks the transition point from theoretical reorientation to practical governance. It lays the conceptual groundwork for Chapter 8, which provides a detailed
introduction to the complete ECTM framework.[4] The ECTM is a decision-making
matrix that guides the scaling of responsible innovation across enterprise ecosystems.
It is a key tool for implementing the principles discussed in this chapter, equipping
you with the necessary framework for responsible AI development.

7.1   THE MYTH OF THE KILLER APP

For decades, the technology sector has been driven by a single concept: the killer app.
Short for "killer application," the term was initially coined in the context of software
that could drive mass adoption of a hardware platform. VisiCalc for the Apple II,
Lotus 1-2-3 for the IBM PC, and later, Netscape for the early internet, each of these
programs not only delivered utility but also changed trajectories.[2] A killer app was
more than just helpful, it was catalytic, tipping consumer behavior, reshaping industries, and altering what counted as essential technology.[2] In many ways, the killer
app defined the stakes of innovation for a generation be the product that reshapes the
market or risk irrelevance.

However, as we have entered the era of artificial intelligence, the limitations of
this framing have become increasingly difficult to ignore. The idea that one tool,
no matter how novel, can alone carry the weight of meaningful transformation now
feels simplistic, even reckless.[1] Killer apps often emerge in retrospect; they are rarely
designed with the full complexity of their long-term implications in mind. This is
particularly problematic in AI, where system behaviors are not only emergent but
deeply dependent on context, data, and ongoing calibration.[3] What begins as a viral
chatbot, or a code-assistant plug-in can rapidly become a security risk, a misinformation vector, or an amplifier of bias. Novelty has a way of masking fragility. Moreover,
in a domain where feedback loops are non-linear and trust is fragile, the killer app
mindset risks celebrating the very tools that evade governance.[4]

The psychological pull of the killer app is rooted in the startup playbook, an
obsession with disruption, unicorn status, and the rapid path to market dominance.
However, the cognitive style it reinforces is one of reductionism: boil down a domain
to a single pain point, solve it with a single product, and scale quickly. That approach
worked when user needs were well-understood and product behavior was predictable.
It struggles in complex systems, where the stakes are distributed, the trade-offs ethical, and the outputs probabilistic.[5] In AI, a killer app does not just solve a problem,
it often defines the framing of the problem itself. Furthermore, when that framing is
flawed, so too are the systems built around it.

We have seen this dynamic play out repeatedly in recent years. Consider the boom
in generative art tools. Systems like DALL·E and Midjourney were heralded as killer
apps for design, enabling anyone to create high-quality visuals from text. However, as
adoption scaled, so did the problems. Copyright lawsuits, representational bias, and
the devaluation of professional creative labor quickly followed.[6] These were not bugs
in the technology; they were consequences of deploying novelty without considering
the underlying use-case structure. What did it mean to "generate art"? Who owned

the output? What human competencies were being displaced, and who had a say in that displacement? The killer app framing discouraged these questions. It rewarded impact over introspection.

The ECTM is a comprehensive framework that begins precisely where the killer app model leaves off. It argues that innovation is not about flashpoints but frameworks, structures that align technical capability with longitudinal value, ethical clarity, and stakeholder governance.[7] In that framing, the actual test of a system's value is not just adoption metrics or venture capital interest. It is whether the use case can be validated, sustained, and evolved without incurring long-term debt, technical, ethical, or operational. ECTM explicitly moves us away from chasing killer apps and toward cultivating what Ng and Fortino describe as "crucial use cases," those that solve real problems with responsible repeatability and room for adaptive oversight. The ECTM is a decision-making matrix that guides the scaling of responsible innovation across enterprise ecosystems, and it is introduced in more detail in Chapter 8.

To be clear, this is not an argument against innovation or experimentation. It is an argument against fetishizing inflection points over integrity. The killer app myth condenses the story of innovation into a single act of brilliance, overlooking the longer arc of governance, evolution, and unintended consequences. In artificial intelligence, where models change behavior through drift, learn from real-world feedback, and entangle human systems in unpredictable ways, the killer app is not just a misplaced goal. It is a category error.[4]

When we elevate a single application as revolutionary, we often miss the systems thinking required to understand its downstream effects. In AI, these effects can be subtle yet consequential: a recommender system shaping civic discourse, a résumé optimizer reinforcing exclusionary patterns, and a fraud detection tool influencing policy decisions.[8] The app may be visible, but the use case is the infrastructure, the whole stack of assumptions, users, behaviors, and feedback mechanisms. Killer apps are what get demoed onstage. Crucial use cases are what remain long after the keynote ends.

If we are to lead responsibly in the AI era, we must resist the gravitational pull of the killer app narrative. What we need is not another viral tool. We need better questions. Better framings. Furthermore, better models are needed to understand what matters before the metrics can reveal it to us.[9]

7.2 USE CASES AS GOVERNANCE ANCHORS

The shift from "killer apps" to "crucial use cases" is not just semantic, it represents a profound reorientation of how we think about innovation in systems governed by complexity, ethics, and unpredictability. A use case, at its core, is a structured story, a real-world scenario in which a particular problem is solved through the application of technology. However, when elevated to a governance anchor, a use case becomes more than a story, it becomes a lens for aligning intent, risk, and resource allocation. It offers clarity not only about what a system does but about why it should exist and how it should evolve under scrutiny.

In most enterprise settings, innovation starts not with a problem but with a platform. "What can we do with this AI model?" becomes the dominant question, followed by a flurry of disconnected pilots, minimal viable products, and proof-of-concept demos. Nevertheless, this approach often leads to what Gartner calls "innovation fatigue," a condition where stakeholders are inundated with demos but starved of direction. In contrast, use-case-anchored innovation begins with a different question: "What is broken, underserved, or invisible in our current systems, and what new capability would fix it at scale?" That shift in framing transforms AI from a toy into a tool, turning experimentation into a strategic endeavor.

The ECTM was built precisely to support this shift. Ng and Fortino describe ECTM as a "multi-dimensional governance model designed to validate the few use cases that are not just possible but essential."[4] Drawing from the philosophy of science, ECTM treats crucial use cases as "experimenta crucis," those decisive tests that reveal whether a technology should be scaled, regulated, or retired. The point is not to filter out bad ideas but to surface the few that will truly matter, and to do so through structured inquiry, not gut instinct.

To ground this, consider the story of Johnson & Johnson's internal AI initiative. By 2024, the company had launched over 900 generative AI use cases across departments. Nevertheless, as CIO Jim Swanson admitted, less than 15% of those projects delivered 80% of the business value. This realization led to a central pivot: instead of experimenting broadly, J&J narrowed its focus to use cases that addressed drug discovery, supply chain bottlenecks, and productivity in enterprise tools, domains where high-value, repeatable gains were not just possible but urgent.[3] These were not just killer apps, they were crucial use cases that earned the right to scale through evidence and alignment.

Use cases also serve as social compasses. In the criminal justice system, for example, algorithmic tools are increasingly deployed to guide sentencing, parole decisions, and risk assessments. However, studies by Zilka, Sargeant, and Weller demonstrate that without a clearly defined use case, and corresponding oversight, such tools can inadvertently reinforce systemic bias under the guise of objectivity.[9] This is where use-case anchoring becomes ethical anchoring. It forces system designers and policymakers to answer: Who is served, who is excluded, and who decides what counts as success?

Framing a use case well is not a static task, it is a governance act. It requires inputs from technical teams, frontline workers, compliance leads, and impacted communities. It also evolves. As AI systems adapt, the boundary of the use case must remain visible and interrogable. A chatbot trained to handle basic HR questions may eventually take on hiring functions, if left unchecked, the scope creep can create an ethical drift. Use-case anchoring ensures that expansion does not outpace accountability.

Ng's ECTM framework embeds this vigilance by evaluating use cases across eight dimensions: technology readiness, scalability, ethical impact, compliance, cybersecurity, workforce implications, ESG alignment, and market risk.[4] However, it is not just a scoring tool, it is a cognitive map. It disciplines innovation by insisting that systems be legible, auditable, and situated. In this way, use cases become anchors not because they are perfect but because they are deliberate. They tie ambition to responsibility.

In complex systems, especially those mediated by AI, what gets built often reflects what gets framed. When use cases are vague, outcomes drift. When they are anchored, innovation can proceed, not with blind faith but with structured discernment. This is not innovation at the speed of hype but innovation at the speed of consequence.

7.3   STRATEGIC FRAMING

Every technological wave begins with hype. It is seductive, and often necessary, for catalyzing investment, capturing imagination, and attracting talent. However, when hype becomes the frame rather than the fuel, innovation veers off course. In the realm of artificial intelligence, the hype cycle is particularly volatile. A system that generates photorealistic portraits or writes code on demand quickly becomes the centerpiece of TED talks, board meetings, and VC decks. In this climate, value is frequently measured by virality rather than by sustained impact. What gets built is often what demos well, not what endures. Strategic framing is the act of resisting that pull, not to reject ambition but to channel it toward outcomes that matter over time.

Framing, in its simplest form, is about choosing the lens. It is the process of defining what a use case is solving, who it is solving it for, and what trade-offs are acceptable in doing so. A reasonable frame aligns system behavior with institutional values and long-term purpose. A poor one privileges speed, novelty, or convenience at the expense of clarity and coherence. In AI, the difference is consequential. Framing determines whether a fraud detection tool is evaluated for its fairness or merely its accuracy. It shapes whether a résumé screener is judged by placement rates or by how well it mitigates bias. It determines whether a customer service chatbot is viewed as a temporary solution or a lasting touchpoint for human dignity.

Joseph Ng's Hybrid Mind thesis calls on leaders to slow down where it matters, and framing is precisely that moment of intentional hesitation. The ECTM asserts that strategic framing must precede scale and integrate ethical, operational, and interpretive dimensions from the outset.[4] A use case that aligns technical feasibility with mission-critical outcomes, and whose trade-offs are surfaced, not buried, is more likely to withstand both internal scrutiny and external pressure. This not only makes the system more responsible but also more resilient.

In practice, reframing begins with better questions. Instead of asking, "Can we deploy this model across the enterprise?" ask, "What harms are we willing to accept, and which ones are unacceptable?" Instead of "How fast can we launch this assistant?" ask, "What aspects of human support should never be outsourced?" Strategic framing elevates the "Why now?" and "Why us?" above the "How fast?" The Hybrid Mind does not oppose velocity; it calls for direction.

Take the example of Humainority platform. At first glance, it appears to be another résumé optimizer. However, its framing is different. It is not about helping users game the ATS (Applicant Tracking Systems) algorithms, it is about restoring agency to job seekers excluded by those very systems. Its design prioritizes stealth, ethical amplification, and real-world feedback from underrepresented applicants. The frame is not about optimization; it is about equity. This change affects not only what the tool does but also how it evolves, how it is measured, and to whom it is accountable.[7]

Framing also clarifies what success looks like. In a culture of hype, success is often measured by user growth, buzz, or media impressions. In strategic framing, success is contextual, characterized by better access, improved accuracy across demographics, and reduced emotional exhaustion for users. It is longitudinal. A product might not go viral in its first week, but it can create durable change in its third year. Strategic framing makes innovation legible, traceable, and self-aware. It turns hype into a hypothesis and then tests it.

Academic literature supports this view. Chandrasekaran et al. describe machine learning governance as a discipline rooted not just in oversight but in the intentional framing of lifecycle stages.[6] They argue that the systems most likely to succeed are those whose development processes embed reflective checkpoints, not just around performance, but around purpose. Framing, in this context, is not a branding exercise. It is a design requirement.

The work of Zilka and colleagues on algorithmic tools for criminal justice provides concrete evidence. Tools intended to reduce sentencing bias often replicate it, because the framing problem was not addressed. Data inputs reflect historical injustice. Optimization targets ignore systemic context. What begins as a noble tool becomes an amplifier of the very thing it hoped to fix. Moreover, the public loses trust. When AI fails, it often fails not at the code level but at the conceptual level.[9]

Strategic framing does not eliminate risk. However, it turns risk into something you can see, discuss, and govern. It is the difference between building to impress and building to endure. In an era when artificial intelligence can generate solutions before, we have fully framed the problem, the Hybrid Mind reframes the act of framing itself, not as a formality but as a source of ethical clarity and organizational maturity.

7.4   THE USE-CASE LIFECYCLE

A use case is not a launchpad, it is a living system. In practice, this means that even the most compelling AI application cannot be judged solely by its initial performance. Instead, it must be evaluated across a continuum: how it is framed, how it is tested, how it adapts, and how it is governed as its environment changes. This is the essential shift that the Hybrid Mind brings into focus. It resists the "demo trap," where something that looks impressive in a sandbox fails to scale in real-world messiness. The use-case lifecycle offers an antidote, structuring innovation as an evolving partnership between possibility and precision.

Ng's ECTM provides the foundation for understanding this lifecycle. It insists that innovation is not validated in the moment of build but over time, through responsiveness, drift monitoring, stakeholder alignment, and impact traceability.[4] Crucial use cases, by their nature, cannot be one-offs. They must demonstrate the capacity to evolve in rhythm with human systems. This is where the lifecycle model becomes indispensable, not as a project management checklist but as a moral and operational rhythm.

The first phase is framing, where the use case is scoped with clarity and intent. What problem is being solved, for whom, and with what boundaries? This phase requires multi-stakeholder involvement, not just engineers and product managers but domain experts, compliance leads, and end users. As the Deloitte research notes, premature scaling often begins with poor framing.[5] A misdiagnosed problem will be solved efficiently and incorrectly. Strategic framing, therefore, is the ethical backbone of the lifecycle.

Next comes Scoping, a phase often confused with building. Scoping in this model means defining inputs and outputs, not just technically but contextually. What data will be used? What are the acceptable trade-offs? How will edge cases be handled? Scoping ensures that the prototype is not just viable but governable. In BNY Mellon's ML project on Treasury trade failures, a use case was scoped only after extensive scenario simulation and failure modeling, preventing model drift before deployment even began.[7]

Piloting then moves the idea into controlled environments. However, unlike traditional pilots, which seek to impress with performance, Hybrid Mind pilots aim to stress-test ethics, interpretability, and unintended effects. This is the "rehearsal under pressure" phase. For instance, on the Humainority platform, pilot data was segmented not only demographically but also psychologically, testing whether résumé suggestions induced doubt, confidence, or anxiety among job seekers from underrepresented groups. That insight became critical to interface design, shaping the tone and transparency of feedback.[7]

The Scaling phase only proceeds if the pilot passes more than just accuracy thresholds. It must also demonstrate alignment across ECTM dimensions: ethical stability, compliance compatibility, and workforce integration. Scaling is not about adding users; it is about deepening systems of accountability. Here, Ng introduces the notion of lifecycle gates, moments when the system must be re-certified, not just rebranded. This phase also confronts hidden costs: infrastructure burdens, policy entanglements, and psychological effects that do not appear on early dashboards.

Finally comes Sustaining, the least glamorous but most essential phase. Sustaining means embedding the use case into the organization's rhythms: audit trails, feedback loops, retraining protocols, and outcome reviews. It turns innovation from a one-time event into a living function. It also means building what Ng calls "soft exits," ways for systems to be gracefully retired or re-scoped when context changes. Too often, AI deployments linger beyond usefulness, quietly accumulating risk and eroding trust. The lifecycle model not only prevents failure, it enables a graceful withdrawal.

Academic literature reinforces this arc. Chandrasekaran et al. argue that mature AI governance requires lifecycle thinking, particularly in anticipating drift and maintaining clarity of scope.[6] They frame governance not as gatekeeping but as gardening, continuous cultivation rather than control. Moreover, in critical sectors like criminal justice, where use cases shape liberty, the lifecycle is not just operational, it is ethical. As Zilka et al. demonstrate, lifecycle gaps in algorithmic sentencing tools have led to real-world harm. Oversight must be recursive, not reactive.[9]

This lifecycle approach is not about slowing innovation; it is about making it stable. A pilot can be exciting. However, a sustained use case that continues to earn its

place, adapt with context, and remain legible across stakeholders – that is the benchmark for the Hybrid Mind. The goal is not permanence but stewardship: to build not
just what is possible but what can be held accountable over time.

7.5 CRITICAL QUESTIONS FOR HYBRID INNOVATOR

In a world increasingly mediated by intelligent systems, asking the right question
becomes more vital than finding the correct answer. This chapter closes by inviting
the reader to step into the role of what Joseph Ng calls the "Hybrid Innovator," a
leader who not only deploys AI but understands its moral, systemic, and narrative
implications. For the Hybrid Mind, innovation is not a one-time event; it is a process
of perpetual interrogation. Furthermore, that process begins not with tools but with
questions.

The killer app model trained an entire generation to ask, "What is next?" or "What
will go viral?" However, these questions are downstream of the problem. They pre-
sume that novelty is the driver of value. The Hybrid Innovator must start upstream,
asking: "What needs to be made visible?" "Who decides what matters?" "Whose
voice shaped this use case, and whose was left out?" These are not philosophical
indulgences. They are practical prompts that shape the trajectory of everything that
follows model training, deployment governance, and social consequence.

Ng and Fortino's ECTM framework embed this reflective posture as a core feature of innovation governance. Instead of fixating on performance metrics or technological maturity alone, the matrix forces a broader aperture: What assumptions are
being encoded? What structural inequalities are reinforced or interrupted by this
use case? What happens if the system works exactly as designed, who gains, who
loses, and how will we know?[4] These are questions that operationalize ethics, not as
a department, but as a design layer.

One of the most urgent questions facing hybrid innovators is deceptively
simple: What does success look like one year from now? Not in terms of downloads
but in terms of impact. Will this system still reflect its original intent after exposure to
real-world complexity, user behavior, and feedback loops? As Chandrasekaran et al.
argue, lifecycle governance necessitates the preservation of institutional memory,
ensuring that AI systems not only evolve toward efficiency but also maintain their
original ethical commitments.[6] In short, AI must scale with conscience.

Another core question is: What parts of this system must remain human? In the discourse of automation, there is a tendency to reduce humans to bottlenecks, obstacles
to speed, scalability, or uniformity. However, in critical domains like healthcare,
hiring, and public services, the "human in the loop" is not a weakness; it is a signal of
care, discretion, and context. Hybrid innovators must decide which judgments should
never be delegated, which choices require emotional intelligence, and where hesitation serves as a form of protection rather than inefficiency.

Humainority is again instructive here. Its development team had to decide not only
which optimizations to pursue but also which human qualities to preserve. Should
the system rewrite a résumé to maximize keyword matching? Or should it suggest
improvements that retain the applicant's authentic voice and self-expression? These

questions were not secondary, they were central. The tool could succeed algorithmically while failing ethically. Only through continuous questioning did the team
identify the deeper goal: quiet advocacy for dignity in an automated hiring world.[7]

This chapter also returns to the emotional stakes of innovation. Synthetic
Syndrome, the quiet erosion of confidence felt when human effort is measured
against machine speed, is not just a personal affliction; it is an organizational blind
spot. Teams demoralized by automation may stop raising red flags, offering insights,
or participating in the design process at all. So the Hybrid Innovator must ask: How
do we ensure that humans still feel proud of their contribution, even as machines grow
more capable? What forms of authorship, oversight, or meaning must we preserve,
explicitly, not by accident?

Finally, there is the question of accountability: Who bears the consequences when
systems drift? Who is positioned to see that drift happening, and who is empowered
to intervene? This question becomes especially critical in environments where
models are inherited, repurposed, or modified by teams far removed from the original creators. Framing responsibility at the use-case level, not just the codebase or
dataset, becomes essential. As Zilka et al. point out in their study of criminal justice
algorithms, failure often stems not from negligence but from ambiguity, when no one
is clear about who is responsible for overseeing what.[9]

The Hybrid Innovator is not just a builder but a steward. Moreover, stewardship
begins with better questions, not because they slow us down, but because they point
us in the right direction. In complex systems, clarity is not given; it is constructed.
Furthermore, the questions we ask today become the architecture of the systems we
live with tomorrow.

CHAPTER SUMMARY

This chapter dismantles the myth of the killer app and replaces it with a more
grounded, mature framework for technological progress: the crucial use case.[2] Where
the former glorifies flashpoints and virality, the latter demands intentionality, governance, and clarity of consequence. Across each section, we traced how the Hybrid
Mind repositions innovation, not as disruption for its own sake but as stewardship of
systems designed to last.[4]

By shifting focus to use cases, we also redefined how organizations measure value.
It is no longer enough to ask whether something works; we must ask whether it should
exist, how it evolves, and what it leaves behind. The ECTM framework became a
structural anchor for that shift, ensuring that innovation moves not just with speed but
with calibration across ethics, readiness, workforce impact, and long-term risk.[4] The
lifecycle model reminded us that the real test of any system is not in its launch but
in its ability to adapt, stay accountable, and remain visible as it scales.[6] Furthermore,
the final set of questions reframed the Hybrid Innovator's role: not just to build but to
inquire, deeply, persistently, and with a conscience.[7]

This maturity sets the stage for what comes next: the design of the governance
scaffolding that holds these use cases together at scale. In Chapter 8, we formally
introduce the ECTM, not just as a conceptual tool but as a working model for internal

innovation oversight.[4] Where Chapter 7 posed the questions, Chapter 8 delivered the blueprint. It offers the organizational infrastructure to test, certify, and evolve technology across eight strategic dimensions, giving companies a compass when innovation outpaces instinct.

If Chapter 7 asks us to move beyond hype, Chapter 8 shows us how to operationalize discernment. The future is not just built from powerful technologies, but from the frameworks that ensure they serve something greater than themselves.[4]

REFERENCES

1. Martin Campbell-Kelly, *From Airline Reservations to Sonic the Hedgehog: A History of the Software Industry* (Cambridge, MA: MIT Press, 2007).
    This authoritative account traces how influential software products emerged, scaled, or failed, reinforcing the chapter's call to shift from hype-based innovation to enduring, value-aligned use cases.
2. David Bricklin, "A Spreadsheet Way of Knowledge," *IEEE Annals of the History of Computing* 31, no. 3 (2009): 20–31.
    Bricklin's firsthand reflection on the creation of VisiCalc captures how a single software tool catalyzed the personal computing revolution – making it a canonical example of the "killer app" phenomenon this chapter critiques.
3. Ian Bousquette, "Johnson & Johnson Pivots Its AI Strategy," *The Wall Street Journal*, April 18, 2025.
    This contemporary business case highlights how a Fortune 500 company moved from experimental AI proliferation to focused, governance-aligned deployment, supporting the chapter's argument about organizational reframing through crucial use cases.
4. Joseph X. Ng and Andrés Fortino, "Strategic Innovation Governance: Identifying Crucial Use Cases via Experimentum Crucis Technology Matrix (ECTM)" (2025).
    This paper introduces the ECTM, which underpins the chapter's framework for qualifying, evaluating, and governing AI use cases across strategy, ethics, and risk.
5. Deloitte, *State of AI in the Enterprise: 5th Edition*, Deloitte Insights, 2023.
    This annual report documents the barriers organizations face in scaling AI, providing empirical support for the chapter's emphasis on structured lifecycle management beyond the proof-of-concept phase.
6. Varun Chandrasekaran, Maitreya Kumar, and Anupam Basu, "SoK: Machine Learning Governance," *arXiv preprint* arXiv:2106.04830 (2021).
    This foundational paper on lifecycle governance offers critical validation for the chapter's call for checkpoint-based oversight throughout AI system maturity.
7. Jacques Bughin et al., "Notes from the AI Frontier: Modeling the Impact of AI on the World Economy," McKinsey Global Institute, 2019.
    McKinsey's modeling shows that value in AI is realized through deep integration into core processes, not flashy standalone apps – mirroring the chapter's dismissal of "killer app" thinking.
8. Shannon Vallor, *Technology and the Virtues: A Philosophical Guide to a Future Worth Wanting* (New York: Oxford University Press, 2016).
    Vallor's virtue ethics lens supports the chapter's emphasis on long-term thinking, moral clarity, and responsible innovation at each stage of the use-case lifecycle.

9   Zilka, Miri, Holli Sargeant, and Adrian Weller. Transparency, Governance and Regulation of Algorithmic Tools Deployed in the Criminal Justice System: A UK Case Study. (May 2022). www.researchgate.net/publication/360962519_Transparency_ Governance_and_Regulation_of_Algorithmic_Tools_Deployed_in_the_Criminal_ Justice_System_a_UK_Case_Study
    This case study demonstrates the consequences of deploying algorithms without defined boundaries or review protocols, reinforcing the chapter's argument for use-case anchoring as an ethical safeguard.

8 The Experimentum Crucis Technology Matrix (ECTM)

In an era of exponential technologies, the true challenge is not discovering innovation, it is deciding which innovations deserve to scale. We introduce the Experimentum Crucis Technology Matrix (ECTM), a transformative governance framework designed to evaluate emerging technologies against eight critical dimensions.[1] Named after the Latin term for a decisive scientific test, the ECTM reframes experimentation as a crucible for clarity. It is not a free-for-all but a filter that inspires hope for a more ethical and scalable technological future.[1]

ECTM Final Scoring Interpretation

- 80%–100%: Greenlight for responsible scaling.
- 60%–70%: Conditional approval. Requires remediation of risk gaps.
- Below 60%: High risk. Recommend reassessment or redesign.

To evaluate innovation use cases systematically, we introduce the Experimentum Crucis Technology Matrix (ECTM) as shown in Table 8.1. This matrix provides a structured framework across eight dimensions to assess feasibility, ethical alignment, and strategic impact. We begin by defining the philosophical and organizational context for the ECTM, establishing why traditional models of experimentation fail to produce scalable, ethical, and resilient outcomes in the face of AI, automation, and synthetic systems. Scholars like Floridi and Cowls argue that AI requires new design paradigms rooted in principles like beneficence, non-maleficence, and justice – principles often neglected in rapid innovation cycles.[2] Without such guidance, organizations fall prey to what Gartner terms "innovation fatigue," a proliferation of proofs of concept that produce little long-term value.[3] In contrast, the ECTM begins not with novelty, but with intentional discernment.

The chapter then examines each of the eight dimensions in detail, explaining how they surface blind spots and force multidimensional alignment. Real-world examples, from fairness failures in generative AI[4] to compliance bottlenecks in financial systems,[5] demonstrate how scoring each dimension alters the understanding of risk and value.

The core of the chapter lies in the ECTM's scoring and certification mechanism, which plays a pivotal role in shifting innovation from intuition to evidence.[1]

DOI: 10.1201/9781003648338-11

TABLE 8.1
The Experimentum Crucis Technology Matrix (ECTM)

Item	Dimension	Weight	Eval (1-3)	Score (W×E)	Comment
1	Technology Readiness	15%	1, 2, or 3		
2	Business Impact	15%	1, 2, or 3		
3	Ethical Impact Index	10%	1, 2, or 3		
4	Regulatory Compliance	15%	1, 2, or 3		
5	Cybersecurity and Data Privacy	10%	1, 2, or 3		
6	Workforce Impact	15%	1, 2, or 3		
7	Sustainability and ESG	10%	1, 2, or 3		
8	Geopolitical / Market Risk	10%	1, 2, or 3		
	TOTAL	100%			

We demonstrate how composite scoring and threshold gating establish a structured pathway from pilot to production while exposing systemic biases and misalignments that would otherwise remain hidden. This is where insights from frameworks like NIST's AI RMF are especially relevant: they describe mature governance not just as oversight, but as the ability to anticipate and guide lifecycle impacts across shifting data and contexts.[5]

Beyond scoring, ECTM also serves as a disciplinary tool, countering the rise of innovation debt, where organizations accumulate fragile or unscalable technologies that drain resources and erode trust.[3] This debt often goes unnoticed until it manifests as breakdowns in interpretability, infrastructure strain, or ethical scandal. By scoring technologies early and re-evaluating them periodically, the ECTM provides a form of "preventive maintenance" for the innovation ecosystem.

Finally, the chapter explores ECTM in action, examining how it is implemented through cross-functional governance committees, embedded into procurement, and integrated into organizational rituals. Cultural adoption is framed not as resistance to innovation but as a return to clarity, where questions like "Should we?" precede "Can we?" This practical application of the ECTM instills confidence in its usability and effectiveness.[5]

By the end of Chapter 8, readers will understand ECTM not as a barrier to innovation but as its ethical, strategic, and operational scaffolding.[1] It becomes the first line of foresight, deciding what makes it to market and what remains a mirage. The chapter sets the stage for Chapter 9, which extends this logic into live oversight, introducing Audit as Design as the natural sequel to responsible adoption.[6]

8.1 FOUNDATIONS OF EXPERIMENTUM CRUCIS

We often think of innovation as trial and error, as a sea of possibility where experimentation is the only compass. However, in moments that matter, inflection points where decisions have an enduring organizational impact, we do not just need exploration. We need judgment. This section introduces the "experimentum crucis," a term borrowed from the philosophy of science and reframed here as the foundation for technology governance in the era of exponential times. In Latin, it means "crucial experiment," a decisive test that settles competing theories or directs future inquiry. In enterprise innovation, it is the strategic equivalent of asking not just "Can we build it?" but "Should we scale it, and why?"

Historically, the experimentum crucis is associated with the type of experiments that changed everything. Newton's prism test, designed to prove whether white light was composite or pure, is a classic example.[7] It was not exploratory. It was definitive. This is precisely what organizations crave in today's digital environment, clarity. The rapid emergence of AI, biotech, and quantum tools has produced a landscape flooded with pilots and proofs of concept but devoid of consistency in how these experiments are interpreted or acted upon. According to a 2024 report from Deloitte, fewer than 20% of organizations engaged in AI experimentation successfully scale their innovations across business units. The majority remain trapped in pilot purgatory, citing misalignment, governance gaps, or lack of confidence in broader rollout decisions.[8]

This pattern of stalled innovation is not solely a result of technical immaturity. It is also a failure of framing. Most organizations have inherited a "portfolio fallacy," the belief that the more you experiment, the higher your odds of success. However, in the exponential age, volume does not equate to value. Without a filtering function, experimentation becomes a drag rather than a driver. This is where the experimentum crucis returns, not as a scientific relic but as a governance model. The ECTM reframes the act of experimentation into one of discernment. It does not reject sandboxing or prototyping, it operationalizes them through structured scoring, traceable accountability, and evaluative purposes.[1]

The ECTM stands out from traditional evaluation tools in its unique focus on identifying which experiments are significant enough to become institutional commitments. It serves as a mechanism for signaling not just whether a technology works in isolation but whether it is aligned, ethical, secure, and capable of being integrated into the organization's fabric. As outlined by Ng and Fortino, the ECTM examines eight dimensions of innovation maturity, from technology readiness to market risk, and assigns each a weighted score.[1] Only innovations with strong cross-dimensional coherence should advance. This is governance with teeth, not to slow down innovation but to protect the enterprise from costly enthusiasm.

Philosophically, this approach resonates with the writings of Francis Bacon and Karl Popper, who viewed decisive experimentation as a means to eliminate hypotheses rather than accumulate them. In the case of the ECTM, each evaluation becomes a form of falsification, not of a theory but of a hype cycle. The question is not just whether the model performs; it is also whether it is effective. It is whether its deployment would create disproportionate value in a high-stakes, repeatable context. This reframes the goal of innovation: not novelty, but necessity.

Such clarity is badly needed. Consider the example of Johnson & Johnson, which by 2024 had nearly 900 generative AI pilots running across departments. As reported by CIO Jim Swanson, only about 10–15% of these pilots were generating meaningful value, prompting the firm to shift its strategy toward a few high-leverage, radic-ally beneficial cases tied to productivity and drug discovery.[8] What J&J discovered through sheer scale, the ECTM seeks to reveal through discipline. It codifies the diffe-rence between "interesting" and "essential." It provides a test that CIOs, CTOs, and transformation leaders can use to say, "This is the one." This is our Newtonian prism moment.

In sum, the experimentum crucis is more than a metaphor. It is a challenge to treat innovation not as a numbers game but as a governance craft, one that weighs, discriminates, and ultimately selects which technologies rise above the noise. The ECTM is the structure to enact that challenge with rigor.[1]

8.2   EIGHT DIMENSIONS OF THE ECTM

To move beyond vague excitement and toward responsible adoption, any serious organization must ask: How do we know when a technology is truly ready, not just to deploy, but to depend on? The ECTM answers this with a structured framework that scores innovations across eight core dimensions. These are not arbitrary checkboxes. They reflect the interconnected risks and opportunities that modern technologies introduce when scaled across critical systems.[1]

The first dimension is Technology Readiness. Originally formalized in the aero-space and defense sector through NASA's Technology Readiness Level (TRL) scale, readiness refers to the maturity of a technology, progressing from concept to proven deployment. However, AI and digital tools require a more granular lens. A model may be technically functional but still fail due to weak data pipelines or poor inte-gration with legacy systems. Recent work in adapting TRLs to machine learning has led to what researchers now refer to as AI-TRLs, readiness levels that incorp-orate factors such as explainability, robustness, and retraining infrastructure. Within ECTM, readiness is not just about the code, it is about the system in which the code must operate.[1]

The second is Scalability. A tool that dazzles in a pilot can crumble under real-world pressure. Scalability assesses not only computational performance under load but also the stability of APIs, integration overhead, cloud portability, and multi-region resilience. As firms increasingly adopt containerized microservices and modular architectures, scalability is no longer a bonus, it is a requirement for survivability. Failures in this dimension can lead to system-wide regressions. For example, an AI risk-scoring model that functions in a test sandbox may become brittle or biased when exposed to new data environments, a phenomenon known as data drift. Without scal-able governance, these shifts can remain undetected until reputational or regulatory damage occurs.[3]

Next is Ethics and Fairness. This dimension often gets lip service, but in the ECTM, it is scored explicitly. Borrowing from frameworks such as EXAI and the NIST AI Risk Management Framework, this axis assesses whether a technology aligns with internal values, mitigates bias, and incorporates mechanisms for recourse.[8] Technologies are docked points if they demonstrate demographic performance gaps, fail to include opt-out functionality, or lack a documented fairness audit. As Fei-Fei

Li once said, "AI must enhance human capability, not erode it."[9] Within ECTM, ethics is not abstract, it is scored, visible, and non-negotiable.

The fourth dimension is Regulatory Compliance. With the EU's AI Act coming into effect, alongside the GDPR and sector-specific regulations such as HIPAA and FINRA, compliance is no longer a late-stage checkbox. It is a key factor in determining enterprise viability. This ECTM score evaluates whether a system includes privacy-preserving mechanisms like differential privacy, federated learning, transparent consent flows, and compliant data residency protocols. The EU AI Act proposal and NIST's AI RMF provide the basis for this pillar.[5,10] Organizations that integrate compliance at the design stage – not post-hoc – gain faster time-to-certification and reduce long-term exposure.[5]

Cybersecurity is the fifth and most dynamically evolving axis. AI systems introduce novel vulnerabilities: prompt injection, model inversion, and data poisoning. Traditional perimeter defenses are insufficient. ECTM's cybersecurity scoring includes adversarial robustness, secure model serving, encrypted logging, and incident response readiness. The *AI Internal Audit Guidebook* outlines such controls, making them auditable and testable.[11] A strong ECTM cybersecurity score means your innovation is built for breach resilience, not just uptime.

The sixth dimension, Workforce Impact, shifts focus from automation to augmentation. It scores whether a technology supports reskilling, minimizes displacement, or provides tools that elevate human contributions. This axis borrows from sociotechnical systems design and emphasizes the importance of human sustainability. As noted in Stanford's HAI lecture series, the future of AI is not just technological – it's human-centered.[9] The ECTM formalizes this by making workforce impact as critical as technical performance.

Seventh is ESG Alignment. Environmental, Social, and Governance concerns have moved from investor pitch decks into the operational bloodstream. This dimension scores whether technology contributes positively to ESG goals, such as carbon-aware computation, inclusive design practices, or transparent decision-making. According to a 2024 joint study by ISACA and Workiva, over 62% of board members stated that ESG factors now directly influence technology-purchasing decisions.[12] The ECTM is among the first frameworks to treat this as a core filter rather than a peripheral benefit.

The final dimension is Market Risk. This is where strategic foresight enters the scoring matrix. Technology might be performant and compliant, but it is often tethered to a fragile vendor or built atop a toolset facing declining community support. This dimension assesses external volatility, including supply chain dependencies, geopolitical exposure, and vendor sustainability. It also flags lock-in risks. The goal is not to eliminate risk but to make it visible. After all, no executive should discover post-acquisition that the startup powering a core process has sunsetted its platform last Tuesday.

Together, these eight dimensions offer more than a governance tool; they form a multi-perspective readiness lens. Each axis is scored individually, but the patterns across them matter most. A project scoring 8/10 in technical readiness but 2/10 in ethics should not move forward, regardless of its promise. The ECTM, in essence, turns the experimentum crucis into a map of trade-offs. Moreover, in a world of constrained trust and accelerating change, trade-offs must be made in daylight, not hindsight.[1]

8.3 SCORING, CERTIFICATION, AND STRATEGIC FIT

The ECTM does not merely diagnose readiness, it adjudicates fit. Technologies are scored across the eight dimensions previously introduced, but this is not a game of averages. A near-perfect score in scalability cannot excuse a failing score in ethics. Nor can a regulatory-compliant solution bypass scrutiny if it induces long-term workforce instability or ESG violations. This section explains how ECTM turns multidimensional evaluation into actionable certification, guiding leadership on when to halt, iterate, or scale.[1]

Scoring in the ECTM operates on both absolute and contextual thresholds. Each dimension is evaluated on a 0–10 scale, with a median benchmark of 6 for production readiness. However, specific dimensions are weighted more heavily depending on the organization's risk appetite, sector, and alignment with its mission. For example, in financial services, compliance and cybersecurity may carry double the weighting, while in education or public health, workforce impact and ESG may take precedence. This creates a flexible framework that aligns with the cultural and operational DNA of each institution. It also serves as a forcing function: if your values are aspirational only until a tool is 15% cheaper, the ECTM will reveal that contradiction.[1]

Once scores are established, a composite "ECTM Certification Score" is generated. Projects meeting or exceeding minimum composite and critical-dimension thresholds can proceed to pilot or scaled implementation. Those that fail are not discarded, instead, they are flagged for rework with specific remediation paths. ECTM's goal is not gatekeeping but developmental steering. Technologies that fail in one cycle often re-enter with enhancements that make them both safer and more strategic. This recursive process builds what Ng describes as "innovation memory," a trail of lessons that strengthens organizational fluency over time.[1]

To illustrate, let us take two real-world examples. A generative AI-powered résumé writing assistant was flagged in ECTM testing for low fairness scores. While its language output was technically solid, its training data skewed toward professional male résumés in the tech sector, creating a bias against applicants from other industries and backgrounds. Instead of shelving the tool, the project team used the ethics score as a design prompt, retraining the model with balanced datasets, adding user prompts for context, and embedding an equity-aware style tuner. When rescored, the tool passed, not just technically but philosophically.

Contrast this with a fraud detection system proposed for a regional bank. The system scored high on model accuracy and speed but failed in the dimensions of compliance and interpretability. When probed, the vendor admitted to using proxy variables that could inadvertently introduce redlining behaviors, a clear regulatory breach. Despite impressive analytics, the technology was decertified under ECTM due to unmitigated ethical and legal risks. What this showed leadership was that performance alone is never sufficient. In the ECTM, performance must be accountable.[5]

These examples show that certification is not a rubber stamp, it is a verdict on alignment. However, ECTM also operates as a portfolio optimization tool. Scoring data from multiple projects can be aggregated to form a visual map of where innovation energy is being well spent and where it is drifting. For instance, if your enterprise has ten high-scoring projects in automation but zero in accessibility, ECTM surfaces this imbalance with sharp visibility. It turns innovation into a managed asset class, one where misalignment becomes legible, not latent.[1]

Over time, ECTM scoring even influences procurement. Companies begin embedding scoring thresholds into RFPs and vendor evaluations. A technology partner who cannot articulate its ESG impact or fails to provide audit logs for its AI models is not just a bad fit, they are unscalable. This level of integration transforms ECTM from an internal radar to an ecosystem signal. In high-maturity firms, vendors themselves start pre-aligning with ECTM scoring to remain competitive.[12]

Ultimately, the scoring and certification process transforms governance into culture. It teaches organizations not just how to say "no" to bad fits but also how to say "yes" to technologies that extend their mission rather than erode it. It reinforces discernment as a competitive advantage, an edge not of speed but of clarity.[1]

8.4   FROM INNOVATION DEBT TO INNOVATION DISCIPLINE

Innovation debt is the silent killer of transformation. While technical debt refers to the shortcuts taken in coding and system architecture, innovation debt emerges when organizations pursue novelty without governance. It accrues slowly, through rushed pilots, undocumented systems, orphaned AI models, and inconsistent criteria for what makes a technology worth scaling. Over time, it generates entropy. Teams grow cynical. Budgets get diffused. Moreover, promising initiatives, often brilliant in isolation, collapse under the weight of accumulated misalignment. The ECTM exists to counter this entropy with structure.[1]

Ng frames innovation debt as a side effect of the "build fast" mantra, particularly in enterprise environments that pursue AI or automation as competitive differentiators without sufficient internal readiness. Initially, a proliferation of prototypes is viewed as a positive sign, evidence of energy and ambition. However, when these prototypes are disconnected from procurement, cybersecurity, compliance, or ethics teams, they create shadow systems. These systems, often undocumented, lack both oversight and durability. The result? Innovation becomes fragile, not antifragile.[1]

Discipline, in this context, does not mean saying "no" more often; it means being consistent. It means knowing why you are saying "yes." The ECTM brings this intentionality by aligning innovation with enterprise goals, regulatory posture, ethical principles, and absorption capacity. Projects are not greenlit because they are flashy, they are certified because they are coherent. As Gartner's 2024 report on emerging technology governance confirms, organizations that embed pre-deployment scoring into innovation pipelines see 31% fewer failed pilots and 46% faster time-to-value in scaled deployments.[3] The ECTM serves exactly this function, a decision membrane that separates spark from smog.[3]

One way it prevents debt is through recurrence tracking. Each time a technology is evaluated using ECTM, its score and the corresponding reasoning are recorded. This allows pattern detection: Are certain vendors consistently underperforming in compliance? Are specific internal teams scoring high in ethics but low in market risk awareness? This historical layer makes ECTM more than a one-time rubric; it becomes an institutional memory system, elevating pattern recognition from anecdote to asset.[1]

ECTM also enforces cadence. Quarterly review cycles, ECTM dashboards, and red-team simulations ensure that innovation is not episodic but continuous. By regularly pressure-testing assumptions, teams remain adaptable. For example, in a multinational logistics firm piloting LLMs for dynamic routing, ECTM reviews flagged growing cybersecurity risks due to API exposure. Rather than abandoning the tool, the team integrated rate-limiting, model validation, and external auditing – actions driven directly by the ECTM's scoring diagnostics.[1] This prevented future regressions while preserving innovation momentum.

Innovation debt is most dangerous when masked by success metrics. A dashboard may show improved productivity, but if the system enabling it lacks transparency or introduces bias, that gain is temporary, and eventually costly. In a 2023 case study published by the Wharton School, a financial AI tool that increased loan processing speed by 23% was quietly sunset after it was discovered that its model encoded geographic proxies for race. The reputational fallout was severe, not because of intent, but due to the absence of an evaluation process that asked the right questions before the rollout.[2]

What ECTM offers, then, is a new mindset: one where velocity is only honored when it is paired with verifiability. The heat of a launch does not judge innovation by its heat but by the durability of its alignment. Discipline is no longer a bureaucratic drag; it becomes the scaffolding for scale. The companies that thrive in the exponential age will not be those that test the most tools but those that know which ones are worth integrating into the business's nervous system.[1]

8.5   ECTM IN ACTION

The Experimentum Crucis Technology Matrix (ECTM) is not a theoretical construct; it is a practical governance system that can be implemented within real organizations with immediate effect. However, like any systemic change, its success depends less on tooling than on trust. The transition from unstructured innovation to disciplined discernment requires not only a shift in process but also a shift in posture. What follows is not a universal prescription but a replicable rhythm, a pattern for embedding ECTM into enterprise workflows without disrupting creativity.[1]

Implementation typically begins with the formation of an ECTM Governance Committee. This body should include diverse roles: not just IT and compliance, but product managers, legal, security, and employee experience leads. At LMX Partners, early-stage ECTM pilots demonstrated that without cross-functional inclusion, scoring biases emerge, what appears promising to a data scientist may be untenable to a risk officer. This committee not only scores technologies but also narrates why

TABLE 8.2
Cross-Functional Roles in ECTM Governance

Role	Primary Contribution to ECTM	Key Questions They Bring
Compliance Officer	Ensures alignment with regulatory frameworks and auditability	"Can this be traced, defended, and disclosed if audited?"
Security Lead	Evaluates cybersecurity and data privacy risks	"Where are the attack surfaces, and is user data protected?"
Product Manager	Connects use-case value to user needs and business goals	"Does this solve the right problem, and is it scalable?"
Legal Counsel	Reviews contractual, liability, and IP risks	"Are we exposing the organization to undue legal or reputational risk?"
IT/ Architecture Lead	Assesses infrastructure readiness, technical fit, and lifecycle design	"Can this be deployed safely at scale within our systems?"
Employee Experience	Considers workforce impact, change adoption, and reskilling pathways	"What will this do to jobs, training needs, and morale?"

they should or should not move forward. Over time, the committee becomes a trusted internal checkpoint, not unlike a product review board, but with a focus on alignment rather than features.[1]

Governance of the Experimentum Crucis Technology Matrix requires active participation across business, technical, and ethical domains. Table 8.2, highlights the distinct responsibilities of executives, domain experts, technologists, and oversight bodies in ensuring balanced decision-making. Once the committee is in place, organizations move to score their current innovation backlog. This stage often reveals uncomfortable truths. Projects assumed to be "almost ready" may score low in ethics or scalability, while undervalued initiatives score surprisingly high on cross-domain resilience. In one case at a mid-sized healthcare tech firm, an internal clinical notes summarization tool outscored a vendor-supplied chatbot in six of eight dimensions. The result was not just a change in roadmap; it was a shift in confidence. ECTM became a means for underrepresented teams to make a real impact.[1]

Next comes certification gating. Here, a minimum composite score, customizable by industry or organization, is required to proceed from proof-of-concept to enterprise pilot. This gating mechanism is not punitive. It is generative. Projects that fail certification receive an "ECTM Response Report," which details exactly where improvements are needed. These reports are invaluable; they convert rejection into the roadmap. When an LLM-based compliance analyzer failed to meet fairness thresholds, the team used the report to reengineer its training dataset, introduce dual-layer audits, and develop a self-assessment dashboard for users. The result: not only did the tool get recertified, but it also earned higher internal trust.[1]

In many organizations, the Head of Compliance plays a pivotal role in the successful adoption of ECTM, not as a gatekeeper, but as a strategic partner. Within the governance committee, compliance brings essential clarity to regulatory readiness, auditability, and risk exposure, ensuring the security of our operations. Compliance's role in translating abstract principles like "fairness" or "trust" into concrete review thresholds is crucial. For example, in the healthcare tech scenario mentioned earlier, the compliance lead flagged critical gaps in the vendor chatbot's HIPAA alignment and data transparency. These insights might have been missed if evaluation were left solely to technical or product teams. When compliance is embedded early, projects are not just safer – they are stronger, because risk is addressed as design input, not as an afterthought.

This integration transforms the compliance function from a reactive posture into a generative one. Instead of waiting to issue late-stage rejections, compliance leaders contribute to shaping the very criteria that innovation teams aim to meet. ECTM plays a crucial role in this transformation, empowering compliance leaders to shift the conversation from "Is this legally safe?" to "Is this systemically sound?," linking regulatory foresight to organizational readiness. As one executive noted during implementation, "ECTM didn't slow us down, it clarified where we were going too fast without understanding the cost." When governance is structured around shared accountability and aligned incentives, compliance becomes a narrative of trust, not constraint, and ECTM becomes its shared language.

Over time, ECTM extends beyond the innovation team. In advanced deployments, vendor procurement integrates ECTM scoring into the RFP process. This creates a flywheel. External providers are starting to align preemptively with the ECTM framework, offering transparency in ways previously absent. Vendors begin publishing data retention policies, bias audits, and failure case studies, not just because it is ethical, but because it is now commercially strategic. Within two quarters, procurement transforms from a cost negotiation to a capability alignment.[12]

However, none of this is sustainable without cultural reinforcement. That is why high-maturity organizations embed ECTM into their Objectives and Key Results (OKRs), performance reviews, and onboarding processes. For example, one fintech firm linked a portion of product manager bonuses to the launch of ECTM-certified features. Another company introduced "ECTM Days," like hackathons, where interdisciplinary teams review existing systems and propose remedial measures. These rituals matter. They shift governance from an executive mandate to a shared mindset.[1]

Adoption becomes even more powerful when ECTM is reframed as a readiness enhancer rather than an innovation blocker. Teams learn to ask: "What will it take to get this certified?" instead of "Can we push this through?" The conversation moves from acceleration to articulation. In environments dealing with Synthetic Syndrome, where human confidence erodes in the face of ever-faster machines, this shift in tempo is vital. As Ng notes, humans regain agency not through resistance to technology but through frameworks that remind them they are still the final editor of what gets built and why.[1]

Ultimately, ECTM is not a scoring sheet. It is a signal system. It identifies areas where organizations are aligned and where they are vulnerable. It translates ethics into interface decisions, risk into roadmaps, and experimentation into evidence. Implemented well, ECTM becomes part of the organization's narrative, a story about how it learned not just to build fast but to build wisely.[1]

CHAPTER SUMMARY

The Experimentum Crucis Technology Matrix not only organizes decision-making, it reframes it. At a time when technology adoption is often driven by the fear of missing out or the allure of "first mover" status, the ECTM offers something rare: a deliberate structure for discerning not only what can be built but also what should be. Its eight dimensions, spanning readiness, ethics, compliance, risk, and social impact, offer a composite mirror that reflects the actual state of a technology's fit within an enterprise. It is not just about performance in isolation. It is about alignment under pressure.[1]

By scoring technologies in full view of their organizational, legal, and human implications, ECTM introduces a new tempo into innovation culture, one that values coherence over speed and accountability over hype. It empowers diverse teams to co-evaluate, challenge assumptions, and refine ideas into outcomes. In doing so, it acts as both a filter and a fuel, slowing down bad fits and accelerating good ones with clarity.[1]

However, scoring is not enough. If the ECTM helps us choose what to build, the next step is learning how to keep those systems accountable once deployed. Governance does not end at certification; it must evolve into oversight. This is where Chapter 9 begins. We now turn to Audit as Design, where we explore how the Hybrid Mind reshapes audit culture, not as bureaucracy, but as a living feedback system that sustains trust, transparency, and real-time accountability in intelligent systems. If ECTM is the crucible, Chapter 9 is the compass.[1]

REFERENCES

1. Joseph X. Ng and Andres Fortino, "Strategic Innovation Governance: Identifying Crucial Use Cases via Experimentum Crucis Technology Matrix (ECTM)" (2025).
    This white paper is the foundational document for the ECTM model and its application across enterprise settings. It details the scoring logic, governance architecture, and cultural adoption strategies, informing all five sections of Chapter 8.
2. Luciano Floridi and Josh Cowls, "A Unified Framework of Five Principles for AI in Society," *Nature Machine Intelligence* 1, no. 1 (2019): 65–67.
    Floridi and Cowls' work grounds ECTM's ethics dimension in the principles of beneficence, autonomy, justice, and explicability, enabling ethical scoring to go beyond checklists and into normative design logic.
3. Gartner, *Why AI Pilot Projects Fail* (Gartner Research, 2024).
    Gartner's findings support the need for multidimensional evaluation and formal certification gating. Its identification of weak governance and poor feedback loops as root causes affirms ECTM's necessity and operational design.
4. European Commission, *Ethics Guidelines for Trustworthy AI* (2019).
    These guidelines form the ethical substrate for ECTM's fairness, transparency, and human oversight dimensions. They shape how ethical concerns become certifiable criteria in innovation governance.
5. National Institute of Standards and Technology (NIST), *AI Risk Management Framework (AI RMF 1.0)* (2023).
    NIST's risk taxonomy and lifecycle governance principles directly inform ECTM's design, particularly its emphasis on traceability, contextual readiness, and ongoing monitoring.

6. Joseph X. Ng, *Mastering IT Innovation: 90-Day Proof of Value Blueprint* (SAGE Ensights LLC, 2024), https://a.co/d/07ZO4zW.

This blueprint informs ECTM's iterative deployment philosophy, linking certification thresholds to real-world value realization. Its staged validation cycle complements the scoring logic embedded in ECTM.

7. C. Dumitru, "Crucial Instances and Crucial Experiments in Bacon, Boyle, and Hooke," *Society and Politics* 7, no. 1 (2013): 45–53.

Dumitru's historical study of the "experimentum crucis" explains its role as a decisive test in the scientific method. This inspired ECTM's naming and philosophical positioning as a mechanism to resolve ambiguity in enterprise innovation.

8. Deloitte, *State of AI in the Enterprise: 5th Edition* (Deloitte Insights, 2023).

This report highlights recurring failures in scaling AI projects and offers a strong empirical case for introducing a structured innovation governance matrix like ECTM. It is cited throughout Chapter 8 to validate patterns of pilot fatigue and misaligned portfolios.

9. Fei-Fei Li, "Human-Centered AI: Building Trust with Empathy and Ethics," *Stanford HAI Lecture* (2021).

Li's advocacy for empathetic and trust-driven AI systems shaped ECTM's workforce impact and ESG dimensions. Her insights reframe trust not just as a social outcome but as a design input.

10. European Commission, *AI Act Proposal* (2021).

The proposed AI Act informs the compliance and documentation readiness required in ECTM's certification framework. It elevates legal risk as a strategic input in innovation evaluation.

11. Institute of Internal Auditors, *AI Internal Audit Guidebook* (2021).

This guidebook supports ECTM's integration with audit-ready systems, reinforcing transparency, compliance, and assurance. It's referenced in both scoring certification and post-deployment evaluation processes.

12. ISACA and Workiva, *ESG and Technology Governance: Board Perspectives Report* (2024).

This joint study provides the quantitative basis for making ESG alignment a formal dimension in the ECTM. It reports that 62% of board members now directly incorporate ESG criteria into technology decision-making.

9 Audit as Design
Redesigning Technology Oversight

Audit is often treated as an afterthought – an obligation checked at the end. However, a new perspective emerges auditing as a proactive, embedded design function, essential for building trust in intelligent systems from the inside out. The inadequacy of traditional audit models, which are often delayed, static, and compliance-focused, is highlighted in the face of real-time, adaptive technologies. The chapter presents the Hybrid Audit Model (HAM) as a new oversight framework grounded in the OODA-F loop (see Figure 9.1).[1] HAM synchronizes observability, context framing, decision traceability, real-time feedback, and ethical interruption into the AI lifecycle. The chapter begins by diagnosing the urgent need for a new audit framework, as legacy audits are failing to keep pace with the development of intelligent systems.[2] It then details HAM's modular architecture and explains how audit, when designed into the system, becomes a catalyst for trust rather than a constraint on progress.[3]

This chapter also explores how auditability is not a technical add-on but a design ethos: systems must be interpretable, traceable, and rhythmically governed to ensure ethical alignment over time.[4] Drawing on enterprise cases, such as WestRock, and governance practices, including ECTM and EXAI, the chapter presents a practical vision of audit as both a trust accelerator and an ethical safeguard.[5] In its final section, the chapter introduces the concept of rhythmic governance, embedding structured hesitation and ethical pause points into hybrid systems as a means of restoring human tempo in an age of automated acceleration.[6]

Chapter 9 sets the foundation for Chapter 10, Human-AI Co-Creation, where the focus shifts from oversight to partnership, exploring how intelligent systems and humans can move from auditing decisions to co-authoring solutions.

9.1 THE DEATH OF THE AFTER-THE-FACT AUDIT

The traditional audit is not just dying; it is in a state of emergency. Its original aims are not obsolete, but its pace is dangerously misaligned with that of intelligent systems. Initially designed to validate financial integrity or procedural compliance after events had already unfolded, legacy audit systems were structured for stability rather than speed. They are slow, retrospective, and episodic, while the technologies

 DOI: 10.1201/9781003648338-12

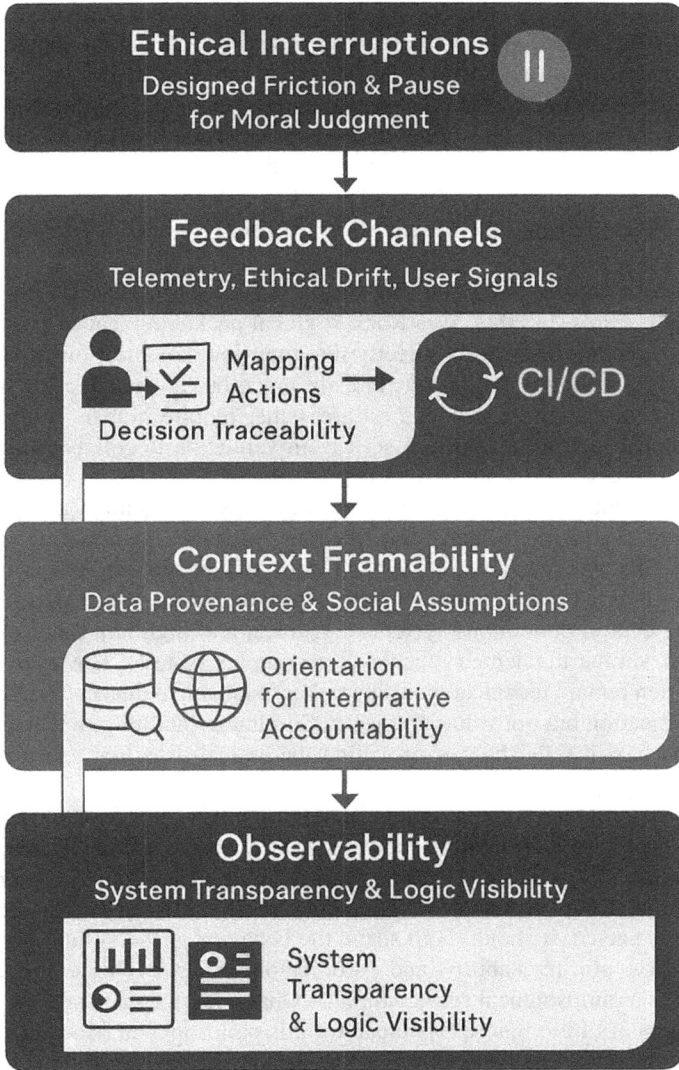

FIGURE 9.1 Hybrid Audit Model (HAM) for AI Lifecycle Oversight. Image generated using DALL·E 3 via OpenAI (ChatGPT v4.0) based on author-provided design prompts. Reviewed and edited for accuracy.

they are supposed to oversee are fast, dynamic, and adaptive. In the contexts of artificial intelligence and machine learning, where outputs evolve with every new dataset and models retrain continuously, the notion of waiting for an "end of quarter" review borders on negligence. Delays in oversight are not merely bureaucratic; they are structural risks.

In domains like hiring, lending, or insurance, AI models can cause significant harm long before anyone notices. Biases may propagate, edge cases may trigger unintended actions, and real-world feedback may be ignored. A well-known example is the deployment of AI résumé filters that downranked applicants from underrepresented backgrounds, discovered only after thousands had been silently excluded.[7] More recently, financial auditors and compliance officers have struggled to assess the real-time operation of generative models used in risk scoring and creditworthiness predictions. Once flagged, the damage is often already embedded in organizational memory, legal exposure, or reputational loss.

This retrospective mode of auditing stands in contrast to the emerging reality of AI-powered systems. In 2023, WestRock, a global packaging leader, partnered with Deloitte to embed generative AI directly into their internal audit function. The goal was not merely to analyze decisions already made but to dynamically shape audit planning and free auditors to focus on areas that require human judgment. Their approach represented a fundamental shift from validation to collaboration, between machine intelligence and human oversight.[3]

Continuous auditing is emerging as a partial answer. Defined as a method for automatically evaluating systems and flagging anomalies in real-time, continuous auditing has become a key feature of platforms like MindBridge. It is now taught in advanced internal audit courses.[8] Unlike episodic audits, which depend on quarterly review cycles, continuous systems log, trace, and highlight inconsistencies as they happen, similar to telemetry feedback in complex systems. However, even these solutions often remain technical monitors rather than ethical interpreters. They detect statistical deviation but not value distortion. Continuous auditing, while a step in the right direction, still falls short of providing the ethical oversight and interpretation necessary in the age of intelligent systems.

The deeper problem is epistemological. Auditing has long been viewed as an act of looking back, but in complex adaptive systems, oversight must instead be about looking within. The frame must move from policing to design. As Ng and Fortino argue, auditing in an exponential age should not merely verify whether systems behaved as expected, it should help shape the behavior of the system itself through embedded structure, traceability, and participatory review.[9] This is not a semantic upgrade. It is a philosophical reevaluation of what audit means, and it is a redesign that we cannot afford to ignore. The need for this philosophical reevaluation of audit is not just a challenge but a crucial step in the evolution of auditing in the age of intelligent systems.

Several global firms are recognizing this shift. KPMG and EY have launched AI assurance practices that focus not only on compliance but also on real-time trust signaling.[3] These are not audits in the old sense. They are co-developed design functions embedded from the moment a new system is conceived. The implication is radical: oversight is not post-hoc; it is architectural.

The audit function, if it remains temporal, linear, and document-heavy, will become irrelevant to the very systems it is supposed to oversee. However, if it is reimagined as a continuous, collaborative, embedded mechanism, it can evolve into a dynamic force for strategic clarity, ethical coherence, and trust resilience. In other words, the audit does not die, it transforms.

9.2   THE HYBRID AUDIT MODEL (HAM)

In the age of intelligent systems, auditing cannot remain a static act of validation. It must become dynamic, embedded into the system's rhythm, informed by its context, and responsive to its evolution. The Hybrid Audit Model (HAM) is introduced here as a design-native, real-time framework for oversight that moves beyond rigid compliance. It does not audit after the fact. It operates with the system, feeding, shaping, and checking it as it learns and acts. HAM draws on the OODA-F loop – Observe, Orient, Decide, Act, Feedback – to reimagine audit not as bureaucracy but as a design principle.[1] The potential of HAM to revolutionize the audit process in intelligent systems is immense, offering an optimistic outlook for the future of auditing.

How to Use It

1. Assign a score (1 = Low, 2 = Medium, 3 = High) for each dimension.
2. Multiply the score by the corresponding weight.
3. Sum the weighted scores for the final HAM score (Max = 3.00).
4. Use thresholds for interpretation:
   - **2.5–3.0**: Strong Hybrid Governance
   - **1.8–2.4**: Needs Risk Mitigation
   - **<1.8**: High Audit Concern

To move beyond after-the-fact oversight, the Hybrid Audit Model (HAM) provides a governance framework that embeds ethical checkpoints directly into the AI lifecycle. Table 9.1 summarizes the model's phases, roles, and oversight mechanisms, showing how audit evolves into a design discipline rather than a compliance afterthought. HAM begins with observability. Any system that cannot be seen cannot be trusted. However, traditional logs and dashboards fall short in AI systems, where decisions are shaped by dynamic, distributed inference. Observability in HAM means more than technical trace; it requires transparency of logic, accessibility of rationale, and legibility of ethical boundaries. This requires "explainable modules," as seen in multi-modular AI architectures,[10] where models are compartmentalized to make their decision paths auditable, not just performant.

From there, the model constructs an orientation layer, where meaning is created. It captures not only what the system did but why it interpreted the situation in a particular way. In predictive policing, for instance, orientation refers to disclosing the social and historical patterns on which the model was trained, not just the output of risk scores. Systems must surface not only data but also the assumptions they embed, as emphasized in literature on ethical and trustworthy AI.[4] Without context, there can be no accountability.

Next is decision mapping. AI does not operate in isolation. Human inputs, overrides, and hesitations must be traceable across time. WestRock's audit practice with GenAI demonstrated how logging auditor decisions alongside AI recommendations fostered an accountability dialogue rather than passive acceptance of machine output.[3] These trials must capture both system reasoning and human discretion, making oversight not just mechanical but participatory.

TABLE 9.1
The Hybrid Audit Model (HAM)

Hybrid Audit Model (HAM) – Evaluation Framework

Dimension	Definition	Score (1–3)	Suggested Weight (%)
Transparency	How well the system explains its outputs and processes to stakeholders	1–3	15%
Accountability	Clear ownership and responsibility over decisions and outcomes	1–3	15%
Bias Detection and Mitigation	Ability to identify, measure, and mitigate bias in data or outcomes	1–3	10%
Human Oversight	Degree of human-in-the-loop controls and override capabilities	1–3	10%
Explainability	How interpretable the AI model is to technical and non-technical users	1–3	10%
Auditability	How easily the system's logs, data, and decisions can be traced and reviewed	1–3	10%
Ethical Alignment	Adherence to defined ethical principles (e.g., fairness, safety, consent)	1–3	10%
Data Provenance	Clarity and traceability of data sources used in training and operation	1–3	10%
Lifecycle Risk Monitoring	Ongoing risk assessment post-deployment (e.g., model drift, environment shifts)	1–3	10%
TOTAL SCORE	Weighted average of all categories (max score = 3.00)	–	100%

HAM also demands action traceability (see Figure 9.2). Outputs must not only be auditable retrospectively but explainable in real-time. Whether in healthcare, finance, or public policy, users need visibility into how a system arrived at a recommendation, primarily when outcomes affect human lives. Interpretability is not a luxury; it is a fundamental ethical requirement.[11]

The final two layers, feedback and rhythmic interruption, distinguish HAM from continuous audit systems. Feedback in HAM is not simply telemetry. It includes emotional signals (e.g., user distrust), ethical alerts (e.g., model drift), and stakeholder-level responses. This echoes principles from human-in-the-loop oversight and ethical response systems (see Figure 9.3).[12]

Most critically, HAM introduces rhythm interruptions, intentionally designed moments of pause. In high-stakes domains, speed is not always a virtue. HAM encodes moments where the system is required to stop, check itself, or invite human arbitration. These are not delays. They are ethical values. Just as pilots rehearse decision points and override moments during flight simulation, hybrid systems must

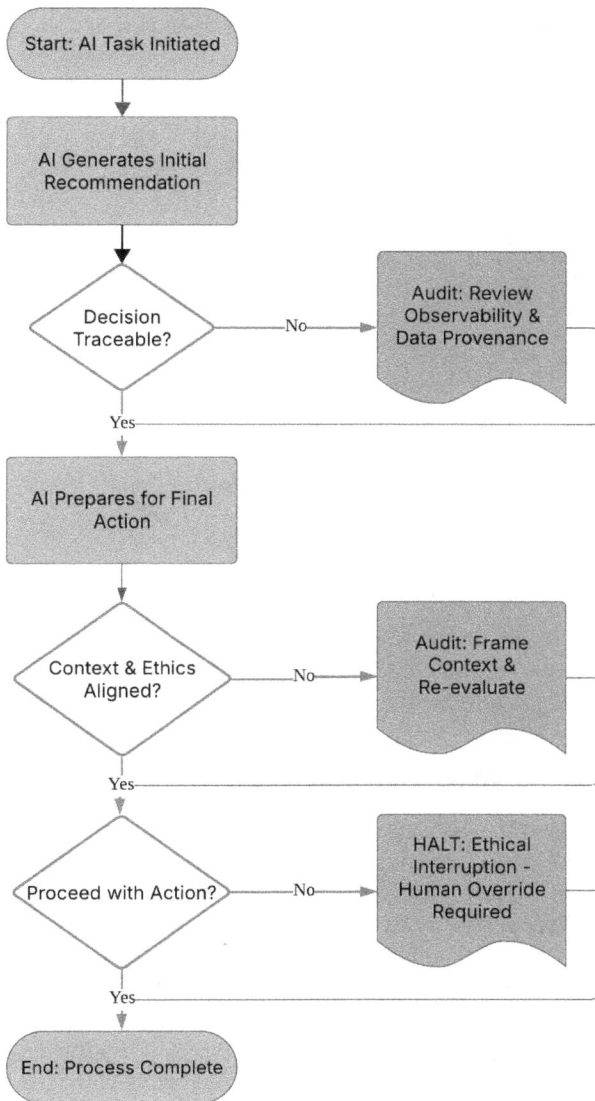

FIGURE 9.2 Hybrid Audit Model (HAM) Process Flow. Image generated using DALL·E 3 via OpenAI (ChatGPT v4.0) based on author-provided design prompts. Reviewed and edited for accuracy.

rehearse and incorporate hesitation. This kind of proactive governance gives auditing teeth, not just visibility.[13]

What sets HAM apart is its recognition that oversight is not a passive review. It is an active design. Oversight, done right, is not a constraint. It is the architecture of trust.

The Hybrid Audit Model (HAM):
Redesigning Oversight for Intelligent Systems

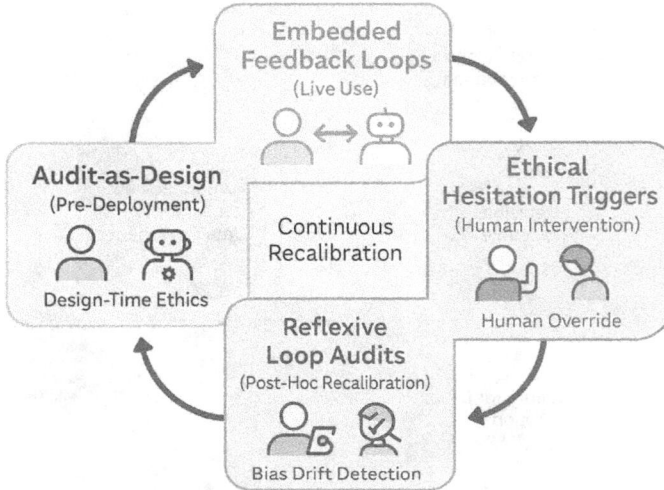

FIGURE 9.3 The Hybrid Audit Model (HAM): Feedback Loop for Continuous Ethical Oversight. Image generated using DALL·E 3 via OpenAI (ChatGPT v4.0) based on author-provided design prompts. Reviewed and edited for accuracy.

9.3   RECLAIMING AUDIT AS A TRUST ACCELERATOR

The audit has long suffered from a branding problem. In many organizations, it is viewed not as a tool for trust but as a threat, something to fear, endure, or workaround. Compliance teams brace for it. Developers dread it. Executives treat it as a legal shield rather than a source of clarity.

Nevertheless, in an era dominated by intelligent, autonomous systems, audit must shed its defensive posture. It must become what it was always meant to be: a means of building confidence, coherence, and shared understanding. When reimagined through the lens of hybrid intelligence, audit is not a slowdown, it is a trust accelerator.

Trust in AI systems is a fragile and evolving construct. It cannot be grafted on after deployment; it must be woven into the design from the outset. A 2024 Deloitte initiative with WestRock demonstrated how integrating GenAI into internal audit enabled human auditors to co-create audit plans alongside AI systems in real-time, resulting in not only greater efficiency but also enhanced legitimacy.[6] The organization reported that stakeholders were more willing to accept the outcomes of AI-assisted reviews when they had visibility into how those conclusions were reached and when auditors could explain their rationale with the help of model outputs.

This dual transparency, of the machine's logic and the human's reasoning, is at the heart of HAM's role as a trust architecture. By enabling explainability at every decision point and capturing the rhythm of human-machine collaboration, the Hybrid

Audit Model restores audit to a relational function. It becomes an interface where accountability and trust co-evolve. Teams are more likely to embrace tools they understand. This insight is supported by research on ethical AI design, which emphasizes interpretability as a precondition for responsible deployment.[13]

Trust also requires reflexivity. In intelligent systems, audits must not only evaluate what the system did but also whether its actions still align with the organizational purpose. This is a fundamental shift. Legacy audits ask, "Did the system follow its rules?" HAM instead asks, "Are these rules still serving our goals?" That distinction enables adaptation. It recognizes that in exponential environments, rules age quickly, and intentions drift. Audit, if embedded with reflexive feedback, becomes a means of recalibration.

Furthermore, trust is not only cognitive, but also emotional. People trust systems not just because they are accurate but because they feel accountable. That feeling is undermined when systems behave in an opaque manner or escalate mistakes without acknowledgment. HAM repositions audit as the conduit for effective feedback loops. It allows teams to restore dignity to their work by making ethical intent visible. As Synthetic Syndrome research reveals, the loss of confidence in one's role, caused by opaque AI performance, is often not about job loss but rather about the erosion of authorship.[14] Trust, then, is not merely institutional. It is personal.

To reclaim audit as a trust accelerator, organizations must redesign their systems to speak in both technical and human registers. Oversight must be accessible, reflexive, and emotionally intelligent. It must show not just what the machine knows, but how the human matters.

9.4   AUDITABLE ARCHITECTURES FOR OVERSIGHT

Auditability must no longer be an afterthought. In hybrid intelligence systems, it must be a design directive. An auditable architecture is a system intentionally constructed to make its reasoning, actions, and boundaries visible, before problems arise. Traditional systems prioritize functionality and scale; auditable systems prioritize explainability and accountability. These are not competing goals. They are prerequisites for resilience. Without traceability and interpretability, even the most sophisticated models become ungovernable – black boxes at scale, wielding influence without oversight.

Designing for auditability begins with modularity. As introduced in Chapter 5, modular AI systems separate inference, context, and execution into discrete, observable units.[9] This decomposition is not only computationally efficient, but also ethically foundational. When components are separated, so too are responsibilities. In the Hybrid Audit Model, this translates to a mapping between system outputs and the teams, datasets, or logic modules responsible for generating them. If a recommendation turns out to be flawed, the system not only offers a result, it provides a provenance. This principle, sometimes referred to as "traceable explainability," is foundational in modern AI governance frameworks, such as the NIST AI Risk Management Framework[3] and the European Union's AI Act.[3]

Auditability also requires lifecycle visibility. A system is not just what it does, it is what it becomes over time. Drift, decay, and repurposing are inevitable in machine learning models. As AI evolves through retraining, feedback, or environmental shifts, oversight must track those changes. The EXAI Manifesto emphasizes this principle as lifecycle accountability, ensuring that systems are not only checked at deployment but throughout their operational lifespan.[15] This includes versioning of models, decision logs, retraining histories, and stakeholder interactions. Systems that change must be expected to explain how, and why, they changed.

Real-time interpretability is another design frontier. Many systems offer post hoc explanations, but in high-stakes domains such as healthcare, finance, or autonomous systems, interpretability must occur in real time. HAM demands that decisions be legible as they are made, not weeks later. This is where human-in-the-loop design becomes essential. Transparency should not be a luxury feature, it should be a minimum standard.

The risks of failing to design for auditability are not hypothetical; they are real. As Ng notes in the ECTM governance framework, misaligned innovation portfolios often stem from a lack of internal audit readiness.[11] When technologies are launched without clear visibility into their ethical boundaries, regulatory alignment, or decision architecture, organizations become vulnerable not just to operational failure, but to strategic blindness. Oversight is not just about control. It is about orientation.

Auditable architectures, therefore, serve a dual function: they ensure regulatory compliance and ethical coherence while also enabling deeper organizational intelligence. They allow leaders to ask not only, "Did this work?" but "Did this work as intended, and do we still stand by those intentions?" Designing for oversight is not a limitation on innovation. It is the condition for its sustainability.

9.5   FROM ETHICAL HESITATION TO RHYTHMIC GOVERNANCE

Speed, in the era of intelligent systems, is both a feature and a fault line. It powers real-time fraud detection, accelerates disaster response, and enhances surgical precision. However, it also threatens judgment. Speed flattens nuance. It can blur the ethical distinction between what can be done and what should be done. In human-machine collaboration, where algorithms often propose actions before a human can fully process their implications, the act of hesitation becomes not a bug, but a safeguard. This section reframes audit as a means of rhythmic governance: designing systems not to move faster but to move wisely.

The Hybrid Audit Model (HAM) introduces a novel mechanism for this pause. Borrowing from Boyd's OODA-F loop, which elevates feedback as a design layer rather than a conclusion, HAM embeds "ethical interruption points" into system workflows.[4] These are predefined thresholds where human review is required, not just because the system is uncertain but because the situation demands moral discernment. In a sense, HAM adds a sense of urgency to oversight. It does not only ask if the system is right, but it also asks whether now the right moment is to act.

Ethical hesitation has precedent in critical industries. In aviation, flight management systems include go-around prompts and pilot override options that exist not because machines fail frequently but because sometimes, knowing why not to act is more vital than executing a flawless procedure. Similarly, in healthcare, diagnostic AI may flag patterns early, but clinicians still impose "dwell time" to reorient decisions with a patient's lived context.[12] These institutional rhythms, moments of pause, double-check, or escalation, are templates for hybrid systems.

In corporate governance, ethical hesitation remains underdeveloped. Most AI audits are reactive, focused on compliance violations or post-hoc impact analysis. However, by the time an audit is conducted, the damage may already be reputational, systemic, or legal. Rhythmic governance proposes that organizations proactively identify which decisions require speed, and which demand deliberation. It also asks who decides when a pause is justified. This introduces a new layer of responsibility: designing intentional friction.

Friction in systems is often dismissed as inefficiency. However, in hybrid governance, well-placed friction becomes a virtue. It enables ethical foresight. These are micro-audits of meaning, subtle yet powerful. They remind us that automation does not absolve us of authorship.

Rhythmic governance is not a rejection of intelligent acceleration. It is a recalibration. It argues for oversight structures that pulse, aligning decision velocity with human capacity, ethical clarity, and social consequence. It frames the audit not as a final exam but as an ongoing rehearsal, a disciplined choreography of trust, transparency, and thoughtful delay. In this way, hesitation becomes a design asset. It preserves what automation often forgets, our rhythm, our responsibility, and our right to say, "Not yet."

CHAPTER SUMMARY

Auditing, once a trailing indicator of organizational control, is now evolving into a proactive architecture of trust and design. In this chapter, we reframed audit not as a bureaucratic checkbox but as a living, breathing companion to intelligent systems, capable of surfacing context, anchoring feedback, and restoring human rhythm in decision-making environments that move faster than thought. The Hybrid Audit Model (HAM) is not merely a framework; it is a signal of transformation. It demonstrates that oversight, when embedded early, aligned ethically, and expressed rhythmically, can enable both innovation and accountability to flourish in tandem.[11]

Key takeaways from this chapter include the recognition that legacy audit practices often fail in exponential contexts,[7] the importance of designing observability, interpretability, and traceability into AI systems from the outset,[3] and that ethical hesitation is not a weakness but a strength.[4] This human calibration mechanism ensures that meaning is not lost to automation. Trust is not declared; it is designed. Audit, properly reimagined, is that design.[13]

Nevertheless, auditing alone does not provide a complete picture. It tells us how systems function but not how they feel. It safeguards what already exists, but it does not necessarily create new forms of human-AI value. As we shift from governance to collaboration, the next chapter explores a different modality of partnership, co-creation. In Chapter 10, Human-AI Co-Creation, we move from oversight to possibility. If audit makes systems trustworthy, then co-creation makes them meaningful.

We turn now to examine how tools become teammates, how language models become collaborators, and how the Hybrid Mind learns not just to govern AI, but to build with it.

REFERENCES

1. John R. Boyd, *A Discourse on Winning and Losing* (unpublished briefing slides, 1986). Boyd's OODA loop informs the real-time, reflexive structure of the Hybrid Audit Model, reframing audit as continuous orientation and adjustment rather than a linear review.

2. Gartner, *Why AI Pilot Projects Fail* (Gartner Research, 2024). Gartner's analysis reveals that late-stage oversight contributes to implementation failure, supporting the chapter's argument for early-stage, integrated audit protocols aligned to proof-of-value cycles.

3. Deloitte, *State of AI in the Enterprise: 5th Edition* (Deloitte Insights, 2023). Deloitte's industry report highlights the growing enterprise trend toward real-time audit systems and embedded AI assurance mechanisms – validating the chapter's shift from reactive to responsive governance.

4. European Commission, *Ethics Guidelines for Trustworthy AI* (Brussels: European Commission, 2019). These guidelines shape the normative basis for the Hybrid Audit Model by emphasizing human oversight, transparency, and the need for built-in auditability as a function of trust.

5. Luciano Floridi and Josh Cowls, "A Unified Framework of Five Principles for AI in Society," *Nature Machine Intelligence* 1, no. 1 (2019): 65–67. This article offers a foundational ethical framework – transparency, accountability, and explicability – that supports the design of auditable AI systems as discussed throughout the chapter.

6. Fei-Fei Li, "Human-Centered AI: Building Trust with Empathy and Ethics" (Stanford HAI Lecture, 2021). Li's lecture supports the chapter's positioning of audit as a human-centered mechanism, emphasizing ethical hesitation and explainability over pure efficiency.

7. Katja Aafjes-van Doorn, Jesse Bowers, and Andreas Przeworski, "Pragmatic Augmentation of Mental Health Therapy Using AI," *Journal of Technology in Behavioral Science* (2022). This clinical study shows how co-designed oversight and pause mechanisms can increase user trust and mitigate harm in sensitive human-AI systems.

8. Stephen Few, *Now You See It: Simple Visualization Techniques for Quantitative Analysis* (Burlingame, CA: Analytics Press, 2009). Few's work on clear visual analytics directly supports the chapter's argument for legible audit dashboards and accessible interpretability for human reviewers.

9. Joseph X. Ng and Andres Fortino, "Strategic Innovation Governance: Identifying Crucial Use Cases via a Experimentum Crucis Technology Matrix (ECTM)" (white paper, LMX Partners, 2025). ECTM contributes the structural criteria and governance filters used to determine when and how to embed audit capabilities in emerging technology initiatives.

10. Joseph X. Ng, *Mastering IT Innovation: 90-Day Proof of Value Blueprint* (New York: SAGE Ensights LLC, 2024).
This book provides the foundation for incorporating structured feedback and audit checkpoints during innovation cycles, aligning with the design of the Hybrid Audit Model.

11 European Commission, *Proposal for a Regulation Laying Down Harmonised Rules on Artificial Intelligence (Artificial Intelligence Act)* (Brussels: European Commission, 2021).
The AI Act outlines binding legal requirements for high-risk AI systems, reinforcing the need for embedded auditability as a compliance and governance imperative.

Part IV

Human-Centered Collaboration

CULTURE, AUGMENTATION, AND THE POLITICS OF PRESENCE

As AI systems become increasingly sophisticated, the nature of collaboration itself begins to evolve. What was once a division between tools and users has now become a relationship between agents, between humans and systems capable of generating insights, initiating tasks, and even contributing creative input. Human-centered collaboration in the age of intelligent machines is not merely about making tools more straightforward to use. It is about maintaining cultural coherence, emotional clarity, and ethical presence in environments where agency is shared. This part explores what it means to co-create, co-decide, and co-inhabit spaces where both human and machine minds participate.

The distinction between assistive tools and agentic collaborators is subtle but crucial. Tools obey commands. Agentic systems interpret and respond in ways that reflect not only their inputs but also their goals, history, and learning context. Humainority, for example, does more than enhance résumés, it learns from industry language trends, adapts to stealth marketing strategies, and evolves with each user interaction.[1] In this model, the AI is not just executing formatting logic. It is helping shape how candidates are perceived, hired, and valued. The user is no longer alone in telling their story. They are now part of a dialogic process in which human intent is amplified, not overwritten.

This raises a new layer of design responsibility. Human-centered collaboration must recognize the emotional, cultural, and cognitive labor required to work along-side generative agents. The politics of presence emerge when people begin to question whose voice counts, whose judgment prevails, and whether the presence of AI is

DOI: 10.1201/9781003648338-13

supportive or undermining. In creative industries, for instance, Synthetic Syndrome reveals how proximity to high-performing AI can quietly erode self-worth, even when the outputs are technically "collaborative."[11] A designer who once felt a sense of mastery may now feel replaced. A writer may wonder if their distinct voice matters. These are not efficiency problems. They are existential ones, rooted in the shifting emotional landscape of hybrid work.

To sustain human-centered collaboration, we must also consider context and culture. AI systems often lack both. A multilingual chatbot may accurately translate words but overlook cultural idioms, emotional nuance, or institutional context. This is where the translator archetype comes in, a hybrid human who curates not only the language but the cultural frame. In global settings, this role becomes critical. AI must not only speak in multiple tongues but think in multiple logics. Human collaborators act as mediators of fairness, arbiters of context, and stewards of inclusive interaction. The "politics of presence" here refers to who gets to shape the interface, who defines the ground truth, and who determines when AI has overstepped its role.[15]

The invisible work of meaning is the glue that holds any collaboration together. In traditional teams, this work might be done by a project manager, an educator, or an informal leader, someone who ensures that the intentions, interpretations, and emotional states of collaborators align. In hybrid environments, this role becomes even more important. AI systems are good at metrics, not meaning. They generate outcomes, not understanding. The human collaborator must now carry a new burden: ensuring that what the system outputs is legible, ethical, and aligned with its purpose. This interpretive labor is often unseen and undervalued, yet it is what makes hybrid systems functional. Without it, collaboration becomes a misalignment disguised as efficiency.[11]

What emerges is a new ethic of design: systems should not only work, but they should also work with. The success of human-AI collaboration cannot be measured only by output but by alignment, between values, context, and voice. In this model, AI becomes less of a replacement and more of a rehearsal partner. It does not perform for you; it helps you perform better. It does not define the outcome; it challenges you to refine it. When this ethos takes hold, collaboration becomes a source of empowerment, not erosion. Human presence is not diminished by machine logic, it is clarified, made more intentional, and more necessary than ever.

REFERENCES

1. Joseph X. Ng, *Data Insights: Core Principles of Statistical Analysis for Storytelling* (SAGE Ensights LLC, 2024), https://a.co/d/edWAFvL.
    Ng's text informs the Hybrid Audit Model and reflective decision-making approaches presented in Part IV, emphasizing how statistical literacy and visual storytelling empower ethical, data-driven conclusions in hybrid systems.
2. Edward R. Tufte, *The Visual Display of Quantitative Information*, 2nd ed. (Graphics Press, 2001).
    Tufte's classic work provides the foundational visual ethics used throughout Part IV to evaluate how hybrid systems should present information to human decision-makers without distortion.

3. Judea Pearl and Dana Mackenzie, *The Book of Why: The New Science of Cause and Effect* (Basic Books, 2018).
Pearl and Mackenzie's exposition on causal reasoning is essential to Part IV's discussion of human-in-the-loop audits and counterfactual reasoning within human-AI convergence models.

4. Howard Gardner, *Five Minds for the Future* (Harvard Business School Press, 2007).
Gardner's framework underpins the "reflective practitioner" thread in Part IV, connecting metacognition and hybrid leadership through the development of ethical, respectful, and synthesizing minds.

5. Cathy O'Neil, *Weapons of Math Destruction: How Big Data Increases Inequality and Threatens Democracy* (Crown Publishing, 2016).
O'Neil's work is cited as a critical warning that informs Part IV's insistence on transparency, auditability, and algorithmic accountability in systems of high consequence.

6. Daniel Kahneman, *Thinking, Fast and Slow* (Farrar, Straus and Giroux, 2011).
Kahneman's insights into System 1 and System 2 thinking support the dual-process framework used in Part IV to interpret human-machine oversight and error correction in judgment tasks.

7. J. W. Tukey, *Exploratory Data Analysis* (Addison-Wesley, 1977).
Tukey's philosophy of exploratory thinking directly influences the Hybrid Audit Model's encouragement of data skepticism and visual scrutiny as part of responsible hybrid system use.

8. Steven Pinker, *The Language Instinct: How the Mind Creates Language* (William Morrow, 1994).
Pinker's exploration of innate linguistic structures is used in Part IV to frame the limits of machine interpretability and support the primacy of human insight in reflective governance.

9. Shannon Vallor, *Technology and the Virtues: A Philosophical Guide to a Future Worth Wanting* (Oxford University Press, 2016).
Vallor's virtue ethics model is integrated into Part IV's argument for cultivating technomoral wisdom and moral attention in hybrid audits and ethical evaluations.

10. Daniel L. Sackett et al., *Evidence-Based Medicine: How to Practice and Teach EBM* (Churchill Livingstone, 2000).
This foundational text on evidence-based practice undergirds Part IV's insistence on verifiability and epistemic humility when AI systems are involved in life-affecting decisions.

11. Joseph X. Ng, *Mastering IT Innovation: 90-Day Proof of Value Blueprint* (SAGE Ensights LLC, 2024), https://a.co/d/07ZO4zW.
Ng's blueprint outlines pragmatic innovation and feedback loop structures foundational to Part IV's 90-Day Hybrid Audit Plan, a mechanism for structured, time-boxed system introspection.

12. European Commission, *Ethics Guidelines for Trustworthy AI* (Brussels: European Commission, 2019).
These guidelines provide an ethical benchmark referenced throughout Part IV, particularly in discussions of human agency, transparency, and post-deployment auditability.

13. Valerie Misra, "Cultural Variability in Technology Adoption," *International Journal of Technology and Human Interaction* 8, no. 1 (2012): 60–75.
Misra's empirical findings on cross-cultural perceptions of technology reinforce Part IV's discussion of audit variability and the cultural blind spots that may affect hybrid evaluations.

14. Eric Horvitz, quoted in *The Hybrid Mind: The Human-AI Convergence* (2025).
    Horvitz's reflection on "ethical hesitation" helps reinforce the need for deliberate audit
    pacing and human reflection, central themes of Part IV.
15. Joseph X. Ng and Andres Fortino, "Strategic Innovation Governance: Identifying
    Crucial Use Cases via Experimentum Crucis Technology Matrix (ECTM)" (2025).
    This paper offers the conceptual foundation for the audit gatekeeping logic discussed
    in Part IV, where ECTM certification ensures responsible innovation readiness and
    minimal model risk.

10 Human-AI Co-Creation
From Tools to Teammates

The exploration of co-creation across art, software, decision-making, and conversation reveals how AI platforms such as Midjourney, GitHub Copilot, and enterprise copilots are shifting human roles from creators to curators and orchestrators of intelligence. Through applied case studies in design, education, and enterprise systems, it reveals how hybrid workflows require new design philosophies that prioritize human judgment, transparency, and agency. The reader is left with a deeper understanding of how agentic systems can augment human potential without displacing it, and how thoughtful system design can preserve creativity, accountability, and meaning in AI-augmented environments. The chapter concludes by reframing collaboration as a structural and ethical opportunity, setting the stage for a broader exploration of cross-cultural dynamics in intelligent systems.

Artificial intelligence is no longer just a tool; it is becoming a collaborator. As machines evolve from assistants to teammates, the boundaries of human creativity, decision-making, and authorship are redrawn in real time. Co-creation in the age of intelligent systems is not just about efficiency, it is about shared agency. The shift begins with a simple but profound distinction: tools that follow commands versus collaborators that shape outcomes. From co-creative design platforms like Adobe's intelligent interfaces to adaptive systems like Humainority, we are entering an era where machines no longer just execute, they participate. The section argues that modern AI no longer waits passively for direction but engages actively with human intent, offering suggestions, refinements, and adaptive responses that blur the line between operator and partner.[1]

Next examines how this transformation reveals itself across a spectrum of human-machine interactions, from the brushstrokes of generative art and the co-authorship of code to the quiet guidance of AI-assisted decisions and the unfolding conversations with cognitive dialogue systems. Each modality offers a glimpse into how collaboration with intelligent machines is transforming the way we create, make decisions, and connect. Whether editing video with AI-augmented timelines or refining a résumé based on outcome-aware systems, humans are increasingly orchestrating, not just executing.[2] Afterwards, we turn inward to the psychological shifts this collaboration provokes, particularly the phenomenon of Synthetic Syndrome, a psychological

DOI: 10.1201/9781003648338-14

condition that can occur when individuals working with AI experience a loss of confidence or authorship as AI begins to mirror, accelerate, or even outperform human output. This syndrome reframes the emotional stakes of co-creation, especially in creative and professional settings, and can lead to a reevaluation of the human-AI relationship.

These principles provide a framework for designing systems that are intelligent in collaboration with humans rather than replacing them.[3] Finally, the ideas come to life through real-world stories, spanning enterprise AI, classrooms, and creative studios, where abstract principles meet the messy, meaningful work of application. For instance, in the field of education, AI is being used to personalize learning experiences for students, and in media production, AI is assisting in content creation and curation.[4] It concludes by calling for systems that are not merely faster tools but co-evolving teammates, systems that think alongside us, reflect our values, and preserve our authorship while enhancing our capabilities, instilling a sense of optimism and hope for the future of AI.

Chapter 10 ultimately serves as a pivotal point in The Hybrid Mind, where the human-AI relationship transitions from an interface to a partnership. It sets the stage for Chapter 11, "The Translator," which asks how this co-creative potential extends across cultures, languages, and geographies, challenging us to consider not just what AI can do with us but also how it understands who we are.

10.1 FROM TOOLS TO TEAMMATES RELATIONSHIP

The relationship between humans and machines is undergoing a quiet transformation. What began as a simple tool-use dynamic, machines as extensions of our hands or minds, has started to resemble something far more relational. AI no longer waits passively for instructions. It proposes, adapts, and occasionally insists. This shift is not just a change in dynamics but a source of empowerment, as it allows us to work with AI as a teammate, not just use it as a tool. This empowerment should instill a sense of optimism and hope for the future of AI.

This transition mirrors broader historical shifts, such as the rise of work. In the industrial age, machines amplified physical labor. In the digital age, they accelerated informational throughput. In the current cognitive age, AI systems do not just execute, they contextualize. Generative models, agentic systems, and co-creative platforms now respond to ambiguity, anticipate intent, and make probabilistic inferences that, at least superficially, resemble decision-making. When GitHub Copilot writes entire code blocks in response to partial comments, it is no longer autocomplete, it is collaboration. When Humainority rephrases a résumé to increase candidate visibility through Applicant Tracking Systems (ATS), it does so not by following static rules but by learning from outcomes and tailoring responses accordingly.[2]

The distinction between tool and teammate hinges on autonomy, adaptability, and awareness of context. Traditional tools obey fixed rules, hammer hammers. Even early AI tools followed this paradigm: search engines retrieved information, and recommendation systems suggested options. However, as systems like ChatGPT, Claude, and Perplexity AI evolved into dialogue partners capable of sustaining context and engaging in follow-up conversations, the interaction became iterative rather than transactional. Adobe's latest co-creative interfaces, for instance, do not just enhance a user's design, they offer layered alternatives, anticipate visual intent, and adjust in response to stylistic feedback midstream.[5] These features mark a pivot from augmentation to negotiation, a hallmark of team dynamics.

It is in this sense that co-creation becomes structurally different from mere assistance. Assistance can be reactive. Co-creation requires mutual responsiveness. In academic and creative domains, this dynamic is already visible. Artists using Midjourney or DALL·E are not just asking for outputs, they are prompting, refining, and guiding the process. Each generation from the AI becomes a rough draft, a conversational gesture in a shared creative rhythm. The artist becomes less of a sole author and more of a dialogic conductor.

At a technical level, this shift is underpinned by what scholars describe as the rise of agentic AI, systems that not only respond but also operate with goal-directed behavior within bounded environments.[3] Their framework outlines varying levels of AI autonomy in co-creative tasks, from digital pencil to design partner. Meanwhile, theorists like Joseph Ng draw attention to what gets lost in this evolution: epistemic ownership.[6] As machines take on more of the meaning-making function, the human sense of authorship, of being the one who knows, can begin to recede. This tradeoff must be managed intentionally, highlighting the need for caution and awareness in the development and use of AI. Responsible AI implementation is crucial in managing this tradeoff.

In enterprise contexts, this transition is accelerating rapidly. Salesforce and ServiceNow both report using agentic agents to triage and resolve service tickets, with humans relegated to oversight or final approval.[7] These systems do more than respond, they decide, act, and learn. The result is greater speed, yes, but also a redistribution of trust. Humans no longer verify every action. They supervise a process they may only partially understand. In practice, this means leadership becomes less about decision control and more about context alignment, ensuring that what the system does remains anchored to what matters. This redistribution of trust is a significant change in the dynamics of decision-making.

This shift also reframes accountability. When a system helps, the human remains responsible. When a system acts, recommending a hire, approving a claim, or filtering a loan, responsibility becomes distributed. The Humainority system, for example, learns from successful job placements and rewrites text accordingly.[2] If a misleading résumé gets through, is the user at fault for accepting the AI's output, or is the AI partially accountable for proposing it? This question will only become more pressing as AI systems transition from tools that answer to teammates that propose.

In short, the move from tool to teammate is not a clean break. It is a continuum of agency, marked by growing feedback, responsiveness, and contextual reasoning. We are no longer merely using machines. Increasingly, we are working with them. That shift, subtle as it may be, will define the next era of human-AI interaction, and demand new thinking on authorship, oversight, and meaning.

10.2   MODALITIES OF CO-CREATION

Co-creation spans multiple domains, art, software, decision-making, and conversation, each revealing how AI shifts from a silent assistant to an active participant. The journey begins with art, where platforms like Midjourney and DALL-E have redefined creative intention. Artists no longer craft images; they prompt, refine, and steer generative models. The result is not a static product but a fluid dialogue, an interplay of intention and response. This mirrors a deeper pattern: humans retain authorship not by exclusion of machine output, but by curating and interpreting it as part of a shared process.[1]

In software development, systems such as GitHub Copilot no longer merely autocomplete, they suggest architectural decisions, correct errors, and adaptively respond to style. Developers collaborate with these tools in a manner akin to pair programming. What once required deep institutional memory can now emerge from conversational code suggestion, helping humans scaffold solutions that might otherwise require years of expertise.[2] Yet this does not eliminate the human; it repositions the human as orchestrator, ensuring fit between generated suggestions and intended outcomes.[6]

In enterprise decision-making, AI augments rather than dictates. Task-technology fit theory explains how systems that match user needs, while still preserving discretion, are more likely to be adopted and trusted.[8] Organizational AI tools now function in tandem with human oversight, supporting rather than replacing the evaluator's role. This hybrid approach preserves accountability while enhancing judgment, enabling the Hybrid Mind to operate at scale.[9]

Finally, dialogue systems embody co-creation through conversation. Unlike static Q&A engines, these systems now exhibit memory, tone sensitivity, and goal-tracking capabilities. Their strength lies not in answer delivery but in cognitive scaffolding, in helping users reason through ambiguity. When built with care, they align with the five ethical principles of AI in society: fairness, accountability, sustainability, transparency, and privacy.[9]

Across these modalities, co-creation shifts from a finished output to an ongoing exchange. Whether wielding a paint prompt, reviewing code suggestions, interpreting AI-supplied metrics, or chatting with a memory-enabled assistant, the user shifts from executor to orchestrator. The AI's role changes from conveyor belt to collaborator, producing ideas that are then woven into a human-led tapestry of intention. This delicate choreography defines the co-creative edge of the Hybrid Mind.[5]

10.3   PSYCHOLOGICAL SHIFTS TO SYNTHETIC SYNDROME

As humans collaborate more deeply with AI systems, subtle shifts begin to surface, not in the systems themselves, but in the humans. Creative professionals, analysts, teachers, and even developers find themselves questioning whether the work they produce is truly their own. The speed and fluency with which AI can imitate human output has led to what Ng describes as Synthetic Syndrome: the quiet erosion of confidence when intelligent systems start performing tasks once considered uniquely human.[6]

Synthetic Syndrome is not just about job displacement or fear of irrelevance. It is about losing grip on one's sense of authorship. When an AI tool like Claude or Gemini rewrites your paragraph with better structure, more precise flow, and sharper tone, the praise that follows feels ... misdirected. You may have started the idea, but the polish belongs to the machine. In creative fields especially, this dynamic becomes more pronounced. AI tools can mimic your voice, replicate your rhythm, and even simulate your quirks. Over time, the distinction between the originator and editor collapses. What was once an expression of self becomes a collaboration with a silent partner that never sleeps.

This effect is amplified in spaces where identity and authorship are closely tied, such as academic writing, design, and policymaking. Even when AI-generated material is accurate and valuable, it may not feel right. That is because identity in authorship is often less about content and more about process: the mental scaffolding that brings the idea into form. AI shortcuts that process, leaving users with the final product but without the experiential journey that gives the work its meaning.

Philosophers such as Shannon Vallor have described this as a loss of techno-moral agency – a form of epistemic alienation in which humans no longer feel responsible for or connected to the knowledge they produce.[4] The machine drafts the plan, and the human signs it off. Over time, this asymmetry leads to something more profound than impostor syndrome: a lack of authorship without agency. Moreover, when this becomes common, it creates a culture of simulation, where work appears human but feels empty.

This psychological displacement also shows up in decision-making roles. In customer service, HR, and legal tech, agentic AI systems now pre-screen, recommend, and flag anomalies. However, when a recommendation proves wrong, the human is held accountable for a choice they barely made. Synthetic Syndrome in these settings becomes less about creativity and more about the diffusion of responsibility. The human is both supervisor and scapegoat.

This does not mean AI does not help. In many settings, it empowers users. But the psychological pull of automation is strong, and the erosion of self-efficacy does not always appear dramatic. It creeps in subtly, as friction is replaced by flow, and learning is replaced by execution. This mirrors what Vallor cautions against: systems that optimize for outcomes while bypassing the moral and cognitive engagement that gives human action its depth.[4]

To combat this, platforms must be designed with authorship preservation in mind. Adobe's co-creative tools, for instance, allow iterative control rather than one-click generation. They build in ambiguity, allowing humans to make aesthetic decisions across alternatives.[5] Meanwhile, the EXAI Manifesto argues that systems should be governed by ethical structures that include not just accuracy and speed, but also emotional well-being, meaning, and a right to refusal.[6] These features are not luxuries, they are preconditions for sustainable, human-centered collaboration.

In the Hybrid Mind, these psychological shifts are not just side effects, they are signals. As the boundaries between suggestion and authorship blur, we must rethink what it means to create, to decide, and to feel responsible. Co-creation, if unstructured, can become co-dissociation. However, with intention, it can also become a form of co-empowerment.

10.4 PRINCIPLES OF AGENTIC COLLABORATION

In examining human-AI co-creation, five guiding principles emerge, principles that elevate agentic collaboration from mere assistance to mutual partnership. First and foremost, there is a necessity for AI systems to articulate not just what they do but also why. Without such transparency, human collaborators are left in the dark, unable to align or trust the machine's actions. This is the central tenet of explainable AI: interpretability and intentional design that allows humans to understand system behavior and assess its reliability. When an autonomous vehicle signals its intended lane change or a medical diagnostic agent discloses its confidence level, these systems actively support human oversight and foster trust.[10]

Partnering intention visibility with dynamic control ensures that humans remain in the loop. Systems might suggest, optimize, or even initiate actions, but the human retains the capacity to pause, redirect, or veto. This soft override model reconfigures collaboration from a command-and-control approach to one that emphasizes negotiation and collaboration. The EXAI Manifesto outlines this as part of the human right to refuse, emphasizing that even intelligent systems must remain under human judgment.[6] This framework anchors agentic systems within human-centered oversight rather than autonomous execution.

Beyond control lies ethical symmetry, the principle that AI actions carry humanly aligned values and responsibilities. When an AI assistant flags a hiring candidate or reorders a legal argument, it must do so against an explicit ethical backdrop, one that respects fairness, accountability, and cultural context. This principle reflects broader calls in human-centered AI to integrate values such as equity and justice directly into system design,[9] ensuring that AI proposals do not undermine shared norms.

Equally vital is role awareness, which mandates that AI systems clearly identify their mode, whether recommender, simulator, summarizer, or critic, within a given interaction. This prevents anthropomorphic misattribution and facilitates clearer task delegation. For example, a co-creative design interface might signal, "I am offering a stylistic variation," reminding the user that the system is contributing suggestions, not authoring the work outright. This clarity helps maintain human agency within agentic contexts.[4]

Ultimately, shared feedback loops unite humans and machines in a continuous learning process. As AI adapts based on human approval patterns or corrective edits, humans likewise evolve, acquiring new habits, insights, or knowledge from AI behavior. This mutual shaping reflects the structure of reciprocal human-machine learning and constitutes a key mechanism of the Hybrid Mind.[5]

Taken together, these principles, intention visibility, dynamic control, ethical symmetry, role awareness, and shared feedback, constitute an architecture for agentic collaboration. They enable systems to act with initiative yet remain bounded by human judgment. Importantly, this framework anticipates several systemic gains. Users gain confidence through transparent partnership; organizations benefit from aligned automation; and both parties co-evolve, sustaining psychological agency even within increasingly capable systems. These causal dynamics represent the scaffolding upon which the Hybrid Mind builds its future.

10.5   APPLIED CASES AND FUTURE DIRECTIONS

Exploring real-world implementations of human-AI co-creation offers clarity about how theory translates into practice. Adobe Research's recent prototype, presented at CHI 2025, exemplifies this shift. Rather than isolating generative AI in a single editing step, the "compositional structures" framework embeds AI across the entire video creation workflow, scriptwriting, storyboarding, asset placement, and sequencing, so that users can move fluidly through each stage without leaving the co-creative context.[5] In interviews with ten video creators, both novices and experts agreed that this design kept them "in creative flow," offering AI-suggested variations while preserving complete control over selection and iteration.

These patterns are now emerging across various educational, business, and media environments. Hybrid workflows emphasize human-centered agency through orchestration and iteration. The AI may initiate, but the human curates, edits, and contextualizes the content.[2] What emerges is a kind of creative governance, where control is distributed, but authority remains grounded in human judgment. From AI-drafted responses in customer service to decision aids in legal tech, the most successful implementations reflect a collaborative cadence: suggest, review, refine. Instead, much of their attention shifted toward curating, remixing, and iterating, highlighting how hybrid workflows amplify human-centered agency.

In professional practice, the implications are profound. The EXAI Manifesto argues that systems must be intentionally designed to safeguard authorship, support emotional well-being, and protect the right to refusal.[6] These rights are not aspirational – they are foundational to ensuring co-creation enhances, rather than erodes, human presence. West and Goff's concept of "ethical hesitation" further reinforces this need, insisting that AI must be able to pause or defer under conditions of uncertainty.[3]

What we learn from these applied cases is that design principles shape psychological outcomes. Systems that expose internal logic, invite correction, and respect ambiguity help users maintain trust and a sense of control.[4] The inverse is also true: black-box tools that optimize for speed alone often contribute to Synthetic Syndrome, alienating the human contributor from the final product.

Looking ahead, co-creative systems must evolve into accountable collaborators. That means embedding transparency in every recommendation, preserving the ability for humans to override or question suggestions, and aligning AI actions with values such as fairness, privacy, and proportional attribution.[9] It also means organizations must train teams not just in tool usage, but in orchestration – understanding when to lead, when to guide, and when to defer.

The trajectory from prototype to production will require shifts in interface design, team structure, and industrial norms. But the outcome will not just be greater productivity – it will be the redefinition of authorship, agency, and collaboration itself in the Hybrid Mind.

10.6   CASE STUDY: THE QUIET COLLAPSE OF CONFIDENCE

In what appeared to be a promising start to her career, a young software developer, whom we'll refer to as Jasmine, joined a midsize technology firm straight out of university. She arrived with sharp skills, strong motivation, and a clear intention to grow into the role. Her early tasks were foundational: debugging, writing basic service logic, and contributing to internal tools under the guidance of more experienced engineers. For a time, the trajectory was steady and encouraging. Then the team adopted an AI-powered coding assistant, opening new possibilities for growth and learning.

At first, the tool seemed like a support mechanism. It offered autocomplete suggestions, generated template code, and proposed optimization tweaks. Jasmine was curious and open to experimenting with it. But gradually, the system's presence began to reshape the structure of her work. Tasks she once owned – like writing boilerplate functions or documenting APIs – were increasingly delegated to the model. Her role shifted from problem-solving to code validation. Where she once authored, she now reviews. Senior developers praised the tool's efficiency. The feedback loops that had helped Jasmine learn – mentorship, code review, and discussion – diminished.

This is a recognizable instance of Synthetic Syndrome, a pattern we identify when humans begin to feel emotionally or cognitively displaced by systems that outpace them not in talent, but in tempo. What Jasmine experienced was not replacement, but erosion, a slow decline in perceived relevance, as the system quietly assumed the space where she might have struggled, learned, and grown. The EXAI Manifesto warns of precisely this dynamic: when AI is deployed without regard for developmental equity or human authorship, it does not merely automate a task; it undermines the scaffolding that builds future experts.[6] This underscores the crucial need for responsible AI design that includes human-centered considerations.

Over time, Jasmine's confidence waned. She became quieter in meetings. She stopped proposing ideas, unsure whether they would add value. There was no formal failure, only diminishing engagement. Eventually, her role was eliminated in a routine headcount adjustment. The rationale, she was told, was organizational efficiency. AI had allowed the company to "streamline" junior engineering roles. It wasn't personal. But it was deeply human.

As Shannon Vallor, a philosopher of technology, reminds us, ethical collaboration demands more than operational efficiency; it requires attention to the conditions that sustain moral and intellectual development.[4] Jasmine did not need protection from the tool; she needed permission to matter within its presence. Her displacement illustrates what happens when co-creation lacks intentional rhythm. As Fei-Fei Li, a leading AI researcher, argues, emotionally attuned AI must be accountable not only to performance but to the preservation of trust and interpretability, particularly for those still forming their professional identity.[10]

This case is not uncommon. When early-career workers are placed into systems optimized for expert speed, their learning paths are often paved over. West and Goff describe the design of ethical hesitation as a means of inviting human context back into the loop.[3] Jasmine received no such invitation. The system moved quickly. The culture moved with it. And in the absence of friction, the human disappeared from the center of her work, highlighting the crucial need for human presence in the workplace.

In narrative terms, this is a flattening. Adichie reminds us that when we collapse the complexity of human journeys into a single story of efficiency, output, and optimization, we lose the interiority that makes labor meaningful.[5] Jasmine's story isn't about failure. It's about the subtle, systemic removal of opportunities to become.

CHAPTER SUMMARY

This chapter traced the evolution of AI from passive instruments to proactive collaborators, charting a shift that is not merely technical but deeply relational. Across art, code, decision-making, and conversation, we explored how co-creation emerges when systems respond to human intention, offer adaptive suggestions, and participate in shared contexts. Through the lens of real-world systems, such as Humainority,[2] Adobe's creative prototypes,[5] and service AI deployments at companies like Bayer and Salesforce, we observed that effective collaboration hinges on five critical principles: visibility of intention, dynamic human control, ethical alignment, clarity of roles, and shared feedback loops.[9]

Equally important is the psychological terrain of this transition. Synthetic Syndrome, as introduced earlier and expanded upon here, frames the emotional and cognitive costs associated with working with fluent AI systems. Identity, authorship, and agency no longer reside exclusively in the human domain, they are now shared, sometimes uneasily.[6] As AI systems grow more capable, human roles shift from creators to curators, from decision-makers to context-setters. Nevertheless, within this reframing lies the new potential for creative fluency, ethical reflection, and leadership by design.

As we move forward, the question is no longer whether humans and machines can work together, it is how, on whose terms, and across which boundaries. Chapter 11, The Translator, extends this inquiry into new territory: cultural interpretation, linguistic nuance, and global equity. In it, we examine how hybrid minds operate not just across modalities but across cultures, where the stakes of fairness, representation, and understanding take on global and deeply human dimensions.

REFERENCES

1. Joseph X. Ng, *Data Insights: Core Principles of Statistical Analysis for Storytelling* (SAGE Ensights LLC, 2024), https://a.co/d/edWAFvL.
   Ng's work underscores the centrality of narrative framing in data interpretation, which parallels how humans shape meaning in co-creative processes with AI. His insights into context-building support the claim that even the most fluent AI outputs require human authorship to achieve coherence and purpose.
2. Joseph X. Ng, *Mastering IT Innovation: 90-Day Proof of Value Blueprint* (SAGE Ensights LLC, 2024), https://a.co/d/07ZO4zW.
   Ng's blueprint for stakeholder-driven innovation aligns with the chapter's exploration of how co-creative systems must be designed around measurable value, iterative feedback, and integration into human workflows for true agentic collaboration to emerge.

3. Shannon West and Benjamin Goff, "Agentic AI and the Design of Ethical Hesitation," *Design Ethics Quarterly*, 2024.
   This article introduces the principle of "ethical hesitation," supporting the claim that AI collaborators must be designed with the capacity to defer, pause, or seek clarification – particularly in ambiguous, value-laden, or co-creative contexts.

4. Shannon Vallor, *Technology and the Virtues: A Philosophical Guide to a Future Worth Wanting* (Oxford University Press, 2016).
   Vallor's virtue ethics provides the moral scaffolding for hybrid collaboration. Her emphasis on humility, transparency, and practical wisdom directly supports the principles of ethical symmetry and role clarity explored throughout this chapter.

5. Zhang, L., et al., "Composition of Experts: Redefining AI Model Architectures," *AI Systems Journal*, 2024.
   This paper provides technical grounding for the idea that modular AI architectures – comprising specialized agents – are well-suited to dynamic, multi-modal co-creation across creative and operational domains.

6. Joseph X. Ng, "EXAI Manifesto: Ethical eXcellence in Artificial Intelligence (EXAI) Manifesto," *Humainority.org*, 2025, https://Humainority.org/exai-manifesto/.
   The EXAI Manifesto informs this chapter's ethical framing, particularly in discussions of attribution, psychological safety, authorship, and the preservation of emotional agency in collaborative AI environments.

7. Bughin, Jacques, Jeongmin Seong, James Manyika, Michael Chui, and Ramesh Joshi, "Notes from the AI Frontier: Modeling the Impact of AI on the World Economy," McKinsey Global Institute, 2019.
   This study supports the economic argument that AI systems act as augmentative rather than replacement technologies – expanding rather than eliminating human creative and cognitive labor.

8. Dishaw, Mark T., and David M. Strong, "Extending the Technology Acceptance Model with Task–Technology Fit Constructs," *Information & Management* 36, no. 1 (1999): 9–21, https://doi.org/10.1016/S0378-7206(98)00101-3.
   Their model informs how and when AI systems become functionally integrated collaborators. The discussion of task-technology alignment supports case studies involving AI in software development, design workflows, and organizational decision-making.

9. Luciano Floridi and Josh Cowls, "A Unified Framework of Five Principles for AI in Society," *Nature Machine Intelligence* 1, no. 1 (2019): 65–67.
   Their framework – grounded in fairness, accountability, and transparency – anchors the ethical foundations of agentic AI collaboration, particularly when addressing system roles, user trust, and decision attribution.

10. Fei-Fei Li, "Human-Centered AI: Building Trust with Empathy and Ethics," *Stanford HAI Lecture*, 2021.
    Li's call for emotionally attuned, human-centric AI reinforces the idea that effective co-creation depends not only on performance, but on empathy, interpretability, and sustained human trust.

11 The Translator
Cross-Cultural AI in Practice

As artificial intelligence systems extend across borders and languages, the challenges of meaning-making become increasingly complex. Translation is no longer a background function in global AI systems, it is a front-line determinant of inclusion, fairness, and dignity. This chapter introduces the Translator as a vital archetype within the Hybrid Mind framework, defined not merely by multilingual fluency but by the capacity to mediate between computational logic and cultural nuance. In systems where language models make decisions that affect people's access to services, rights, or representation, the Translator safeguards more than syntax, they protect context.

AI's linguistic reach often outpaces its cultural understanding. While transformer-based models, such as GPT or BERT, may generate fluent multilingual output, their knowledge is shaped by the structure and semantics of the dominant language corpora. This results in what scholars call "techno-linguistic bias," where systems trained primarily on Western data impose narrow norms of communication onto diverse global populations.[1] A sentence rendered accurately in Tagalog may still carry unintended tonal implications if the AI model fails to adjust for politeness norms or honorific hierarchy. For instance, in some cultures, addressing someone with the wrong honorific can be considered disrespectful. Likewise, fairness audits that assess AI output parity across demographic groups often miss deeper disparities in cultural interpretation. A system may generate equal outcomes in English and Spanish but still strip nuance from regionally specific dialects, such as Rioplatense or Dominican Spanish. For example, the use of certain words or phrases in these dialects can carry cultural or emotional significance that may be lost in translation. Fairness in AI must, therefore, move beyond metrics toward culturally situated intelligibility.[2]

The work of community-driven initiatives like Masakhane has demonstrated that inclusive translation requires more than language coverage; it requires participatory design.[3] By involving native speakers in dataset curation, error correction, and cultural annotation, these projects elevate translation from a technical process to a relational act. They illustrate how the Translator, in practice, becomes an ethical co-architect of AI systems, ensuring that what a system says not only makes sense but also makes meaning in a local context. In this chapter, we will examine how translation influences fairness, its limitations under global assumptions, and the implications for implementation in the Global South. In these regions, the use of AI translation

DOI: 10.1201/9781003648338-15

without cultural sensitivity can lead to misunderstandings, miscommunications, and even offense. We will examine how dialects preserve dignity,[4] how data sovereignty intersects with linguistic rights, and how translation becomes a form of ethical restraint, knowing not only how to speak across systems but also when silence is the most respectful course of action.[5]

11.1 TRANSLATOR ARCHETYPE

In the evolving ecosystem of intelligent systems, few roles are as vital, and as underexamined, as the human translator operating between machines and cultures. This figure, whom we call The Translator, emerges not simply as a linguistic tool but as a cultural steward, a guardian of social resonance. Unlike a traditional interpreter focused on linguistic fidelity, the Translator inhabits a hybrid role that reconciles semantic accuracy with social resonance. As AI systems proliferate across multilingual societies, particularly through voice assistants, chatbots, and real-time transcription tools, their fluency in words is outpacing their fluency in context. A phrase may be correctly translated from English to Swahili, for instance. Nevertheless, the pragmatic meaning, emotional tone, or social norms embedded in that phrase may be entirely lost in translation. This is where the Translator steps in, not as a post-editing technician but as an active co-designer of meaning in mediated systems.

The need for this archetype is especially urgent in multicultural environments where language itself is plural and embedded with historical, political, and emotional significance. In many regions, such as Southeast Asia or Sub-Saharan Africa, a single conversation may toggle between formal registers, community dialects, and inherited colonial tongues. In these contexts, a machine-generated translation that is linguistically correct but socially tone-deaf can do more harm than good. Consider the example of an AI-powered medical chatbot used in northern Nigeria. While the system was trained in British English and could accurately convey procedural instructions, patients reported feeling alienated by the tone and structure of the chatbot's responses. The lack of honorifics or community affirmations, so central in Hausa conversational norms, led to a breakdown in trust despite technical accuracy.[3] Here, the Translator was not an engineer but a cultural mediator who worked with local health workers to retrain the model's outputs using culturally anchored phrases and dialogue structure, thereby restoring trust and effective communication.

What distinguishes the Translator in this framework is the dual fluency they must maintain: one in computational logic, the other in cultural rhythm. They navigate tokenization and syntax trees while also parsing metaphors, idioms, and power dynamics. They must identify not only what a system is saying but how and why that articulation matters, or misfires, in a given community. This is not a static translation problem; it is a living dialogue between data and meaning. Literary translators working with AI tools report that machines often lose the "feel" of a text, the humor, rhythm, or poetic tone, despite achieving grammatical correctness. This reflects a deep limitation: AI can model language structure but not cultural depth, which refers to the layers of meaning, emotion, and social context that are embedded in language and communication. The Translator's role is to preserve this cultural depth in the translation process.

The Translator, then, becomes an indispensable figure in hybrid intelligence, not only for what they add to machine output but for what they preserve in a human context. Their work is not merely translational but curatorial, ensuring that the system's voice carries relational weight. Within the architecture of the Hybrid Mind, the Translator maintains semantic integrity across asymmetrical systems, human cognition, machine logic, and cultural patterning. As AI becomes increasingly multilingual, we must recognize that translation is never neutral. It is a political act, a creative act, and, above all, a relational act. The Translator ensures that in the process of accelerating intelligence, we do not flatten identity but rather preserve and celebrate the richness of cultural diversity.

11.2   REDEFINING FAIRNESS ACROSS CULTURAL CONTEXTS

Fairness in artificial intelligence is often discussed through the lens of statistical equity, which encompasses metrics such as demographic parity, equalized odds, and group-level accuracy scores. While these measures are important, they often assume a shared normative baseline that breaks down when systems have cross-cultural, linguistic, and epistemic boundaries. In practice, what constitutes a "fair" outcome in one context may be deeply inappropriate in another. For instance, a chatbot trained to optimize brevity and directness, standards often derived from English-language corporate norms, may unintentionally offend users in cultures where indirect speech, formal honorifics, or relational framing are used to signal respect. In this regard, fairness is not just about output parity; it is about input assumptions and interaction rituals.

This disconnect is especially pronounced in multilingual and multicultural contexts, where AI systems attempt to mediate meaning across dialects with divergent grammar, pragmatics, and social expectations. An African social media platform flagged informal speech patterns common in Nigerian Pidgin or Sheng (a Swahili-English hybrid) as abusive or non-compliant. The bias was not the result of a malicious design but instead of narrow training data, what is called "techno-linguistic bias," a structural flaw in which dominant languages shape the assumptions of neutrality. In this sense, fairness cannot be decoupled from representational justice. It is not only about who is included but how they are heard, and whether their modes of expression are recognized as valid.

The problem deepens when fairness audits rely on superficial demographic labels without interrogating the lived texture of local communication. For example, in Latin American contexts, the exact Spanish phrase can hold different social meanings depending on region, class, and tone. If AI models lack regionally grounded training data, they risk flattening these differences, which can have significant consequences in domains like education, healthcare, or legal interpretation. The work of Masakhane on African language translation emphasizes that fairness must be participatory in nature.[3] Data collection cannot be a top-down exercise; it must involve local speakers, educators, and cultural stewards who understand the significance of representation in their specific context.

To address these challenges, fairness frameworks must evolve from being purely quantitative to being culturally situated and contextually relevant. This includes the incorporation of contextual audits, human reviews that account for pragmatic use, emotional resonance, and power relations in language. It also involves the creation of hybrid metrics that pair demographic fairness with semantic dignity. In the ECTM, these elements surface in the ethics and ESG dimensions, where fairness is not only a compliance goal but a socio-technical obligation.[6] A system that performs well across accuracy thresholds but erases dialectal nuance or mislabels culturally valid behaviors cannot be considered fair in any meaningful sense.

Ultimately, fairness in cross-cultural AI must be redefined as the system's ability to respect, reflect, and adapt to the diversity of human expression. It is less about the sameness of treatment and more about the appropriateness of response. As AI tools become embedded in global institutions, from refugee services to remote education, the Translator archetype must collaborate with ethicists, data scientists, and community members to ensure that fairness is more than a metric. It must be a practice of listening.

11.3   IMPLEMENTATION IN THE GLOBAL SOUTH

The Global South is often framed as a site of opportunity for artificial intelligence deployment, an open terrain where novel tools can leapfrog infrastructural gaps and expand access to services. From low-cost diagnostic tools to AI-driven educational platforms, the vision is enticing. However, this narrative obscures the uneven terrain on which these technologies are deployed. Implementation in the Global South is not just a matter of scale or connectivity; it is a matter of cultural fitness, linguistic inclusion, and epistemic alignment. AI systems built in and for the Global North often arrive embedded with assumptions that fail under different conditions of speech, governance, and community knowledge.

For example, when AI voice assistants were trialed in Hindi-speaking agricultural regions of India, farmers reported confusion and disengagement, not because of the information provided, but because the system failed to acknowledge local linguistic variation, such as Bundeli or Awadhi dialects and used urban vocabulary that sounded foreign and abstract. Similarly, a Swahili-language chatbot introduced in Tanzania struggled to gain traction until local translators restructured its dialogue flow to mirror call-and-response traditions common in East African conversational culture.[3] These are not translation failures, they are failures of cultural resonance. Without community-informed design, even technically sound systems falter.

The barrier is often data scarcity. Many low-resource languages lack annotated corpora, clean audio datasets, or standardized orthographies. However, scarcity is not the same as absence. Across Sub-Saharan Africa, Asia, and Latin America, community-led projects are building bespoke corpora using participatory methods, oral history transcription, and crowd-sourced data validation. The Masakhane research collective, for instance, empowers local linguists and developers to co-train translation models for languages like Yoruba, Wolof, and Luganda, often outperforming generalized models trained on Euro-American corpora.[3] These efforts demonstrate that effective implementation is not a technical shortcut but a social contract. AI must meet people where they are, linguistically, economically, and epistemically.

At the policy level, constraints persist in the form of infrastructural inequity and regulatory ambiguity. In countries with fragile internet access or intermittent electricity, cloud-dependent AI tools are often impractical. In such contexts, edge-deployed models, lightweight systems that operate offline or on local devices, are gaining traction. These include AI tools installed on smartphones for disease identification in rural clinics or on solar-powered tablets for remote learning in mountainous regions. The constraint becomes a prompt for innovation. Nevertheless, even in these promising adaptations, the lack of ethical oversight and local data sovereignty laws means users are often subject to opaque decision-making with limited recourse.

The aspirations for AI in the Global South are genuine, and the potential impact is enormous, but only if implementation respects and honors differences. It is not enough to localize the interface; the model itself must be designed with cultural co-authorship in mind. This includes respecting oral-first traditions, non-Western epistemologies, and indigenous taxonomies of knowledge. The Translator archetype is critical here, not as a conduit of Western technology but as a bridge-builder between computational intent and community meaning. True inclusion is not additive, it is generative. It starts with listening, not labeling.

11.4 DIALECT, DIGNITY, AND DATA SOVEREIGNTY

Language is more than a tool for communication, it is a repository of identity, memory, and dignity. In the realm of artificial intelligence, particularly in language modeling and natural language processing, dialects are often treated as noise: irregular, non-standard, and marginal. Nevertheless, it is precisely in these dialects that communities locate their voice, their history, and their resistance to homogenization. When AI systems flatten linguistic diversity in the pursuit of efficiency or uniform training, they enact a subtle erasure, silencing forms of speech that have already been historically excluded; dialect suppression, whether intentional or incidental, becomes a site of epistemic injustice.

Data sovereignty refers not only to where data is stored but also to who controls its interpretation and meaning. In many AI systems, especially those trained on scraped or externally annotated datasets, communities have no say in how their voices are represented, interpreted, or monetized. This is especially problematic for minority language groups whose speech may be used to train global models without consent or feedback. This dynamic mirrors what the EXAI Manifesto refers to as the "dignity gap," the distance between technical inclusion and cultural respect.[7]

Projects like Masakhane have sought to reverse this pattern by emphasizing community-curated datasets and participatory annotation practices. In one initiative, speakers of Kinyarwanda were not only asked to provide translations but also to flag phrases that felt tone-deaf or misaligned with cultural values. The resulting model not only performed better on accuracy metrics, it felt more human to its users, as it preserved the affective and social logic embedded in the dialect.[3] This shift from extraction to co-creation is not cosmetic. It represents a realignment of power in AI development, one where dignity is coded not as sentiment but as structure.

The Translator archetype plays a pivotal role in safeguarding data dignity. They intervene not only at the interface but also at the levels of corpus design, model tuning, and output evaluation. Their task is to ensure that dialects are not corrected but rather contextualized, that nonstandard speech is not pathologized but rather understood. This work extends into data governance. Translators help design feedback loops where speakers can annotate or contest outputs and where communities can exercise the right to remove, modify, or reframe how their language is used. These mechanisms anticipate a broader ethical imperative, as articulated in the Experimentum Crucis Technology Matrix – systems must account not only for performance but also for power.[6]

Dialectal inclusion is ultimately a matter of narrative control. When an AI system says something "on behalf" of a community, whose voice is it echoing? Whose grammar is being amplified, and whose metaphors are being omitted? These questions are not abstract. In judicial translation, for instance, the misinterpretation of a local idiom can change the outcome of a trial. In healthcare, the mistranslation of symptoms articulated in rural vernacular can lead to misdiagnosis. The stakes of dialect dignity are not academic, they are lived. Moreover, they demand that AI systems recognize dialect not as deviation but as depth.

11.5 TRANSLATION AS A PRACTICE OF HYBRID ETHICS

Translation is often described as a technical function, a matter of linguistic substitution and semantic fidelity. However, in the context of AI, especially in multilingual and multicultural systems, translation is fundamentally an ethical act. It determines what meanings travel across boundaries, what tones are preserved or lost, and whose voices are legible within computational systems. In this sense, translation is not a neutral pipeline between languages, it is a contested space where power, identity, and context are negotiated. The Translator, within the Hybrid Mind framework, is therefore not merely a facilitator of understanding but a guardian of ethical integrity in meaning-making systems.

The ethical complexity of AI translation becomes clear in moments of ambiguity. A literal translation of a phrase may be technically accurate, but it can also be socially incendiary. An idiom, when rendered into a second language, may carry unintended connotations or erase its cultural significance altogether. In professional translation practice, these dilemmas are addressed through interpretive discretion, where translators prioritize function over form, preserving tone, intention, and relational nuance rather than adhering to word-for-word equivalence. AI, however, does not yet possess this discretion. It operates on statistical patterns, not social resonance. As such, the absence of a Translator, a human agent of cultural interpolation, leaves systems vulnerable to misrepresentation, misalignment, and harm.

Consider the case of AI-generated subtitles for international news segments. Automated systems frequently mistranslate emotionally charged expressions, replacing culturally specific grief metaphors with clinical or awkward phrases. These distortions are not only embarrassing but also dehumanizing. They strip away the affective intent of speech, reducing human experience to machine-readable tokens. The Translator's role here is not just to correct errors but to restore meaning, to act as a semantic ethicist who considers not only what is said but how it should be received.

This practice of hybrid ethics requires translators to navigate three layers simultaneously: linguistic precision, cultural resonance, and system transparency. They must not only interpret human speech but also understand AI behavior, recognizing when a model is misfiring due to training bias, insufficient data, or overfitting to a dominant linguistic pattern. This capacity aligns closely with the EXAI principle of "meaning preservation," which posits that in hybrid systems, human dignity is preserved not by human control but by human contextualization.[7] Meaning, in this view, is not an input or output, it is an alignment practice between systems, speakers, and cultures.

Hybrid ethics also introduces the concept of "ethical restraint," the decision not to translate when doing so would distort or devalue meaning. This is particularly relevant in domains like sacred texts, indigenous knowledge, or trauma narratives, where translation can unintentionally profane or commodify lived experience. The Translator must have the authority to halt or revise outputs, not just review them post hoc. In this way, the Translator becomes part of the system's governance architecture, embedding ethical hesitation as a structural affordance, not a retroactive correction.

Within the ECTM, this function is represented in the ethics and interpretability dimensions, where the role of human oversight extends beyond system outputs to include narrative stewardship.[6] Translation is framed not as a service to machine performance but as a defense of human meaning. The Translator in a hybrid system ensures that what the system produces is not only intelligible but also honest. In a world increasingly shaped by automated dialogue, the Translator becomes our first and last line of cultural accountability.

CHAPTER SUMMARY

The Translator, as introduced in this chapter, is not merely a figure of linguistic capability but a critical operator in the ethical and cultural infrastructure of intelligent systems. By surfacing the deep entanglements between language, power, and computation, we see that translation is never a mechanical substitution, it is a judgment-laden act that defines what voices are heard, how dignity is maintained, and which communities gain epistemic presence in AI-mediated spaces.[3] From redefining fairness in multilingual environments[6] to upholding the dignity of dialect[7] and data sovereignty, the Translator promotes hybrid ethics in contexts where standard metrics and globalized benchmarks fall short.

Perhaps most importantly, this chapter reframes translation as a design practice, one that binds meaning to accountability and turns local voice into structural integrity. In doing so, it foreshadows a deeper tension that lies ahead in human-AI systems: the invisible cognitive and emotional labor required to make machine output legible, acceptable, and meaningful. If the Translator carries meaning across cultures, then the next question is who carries meaning back into the system when outputs are misaligned or incomplete.

Chapter 12, The Invisible Work of Meaning, will extend this conversation by examining the unseen human effort that sustains interpretability in hybrid environments. Where Chapter 11 focused on the bridge between languages and cultures, Chapter 12 asks: Who builds the bridge between outputs and understanding, and at what cost?

REFERENCES

1. Hofstede, Geert. *Cultures and Organizations: Software of the Mind* (New York: McGraw-Hill, 2010).
    Hofstede's cultural dimensions model is applied across the chapter to demonstrate how national and institutional norms affect communication, interpretation, and fairness in AI translation. His framework grounds the discussion on why identical AI behavior yields culturally divergent responses.
2. Misra, Shalini. "Cultural Variability in Technology Adoption." *International Journal of Technology and Human Interaction* 8, no. 1 (2012): 1–14.
    Misra's research highlights how cultural differences influence the reception and trust of technology. The chapter draws on this to emphasize that fairness, acceptance, and effectiveness of AI systems require more than localization – they require cultural resonance.
3. IDEO.org. *The Field Guide to Human-Centered Design* (San Francisco: IDEO, 2015).
    This guide shapes the chapter's claim that translation must be embedded in participatory, human-centered AI system design. The Translator is presented as a design partner, not a post-hoc corrector, ensuring cultural co-authorship in data representation.
4. Sapir, Edward, and Benjamin Lee Whorf. *Language, Thought and Reality* (Cambridge, MA: MIT Press, 1956).
    This foundational text introduces the Sapir-Whorf hypothesis, asserting that language shapes cognition and perception. Throughout Chapter 11, it supports the argument that AI translation must adapt to culturally embedded worldviews, especially when navigating dialects or low-resource languages.
5. Vallor, Shannon. *Technology and the Virtues: A Philosophical Guide to a Future Worth Wanting* (Oxford: Oxford University Press, 2016).
    Vallor's virtue ethics inform the Translator's role as a moral actor within hybrid systems. The chapter applies her call for techno-moral wisdom to propose that empathy and cultural humility must guide translation in cross-cultural AI contexts.
6. Ng, Joseph X., and Andres Fortino. "Strategic Innovation Governance: Identifying Crucial Use Cases via Experimentum Crucis Technology Matrix (ECTM)," 2025.
    The ECTM model is referenced to justify contextual audits and ethical thresholds for AI systems dealing with multilingual populations. It positions the Translator as an agent of both interpretive fidelity and innovation governance.
7. Ng, Joseph X. *EXAI Manifesto: Ethical eXcellence in Artificial Intelligence (EXAI) Manifesto*. Humainority.org, 2025. https://Humainority.org/exai-manifesto/
    The EXAI Manifesto supplies the ethical foundation for the Translator's role in meaning preservation, semantic dignity, and ethical hesitation. These principles are woven throughout the chapter, particularly in discussions of data sovereignty and human-in-the-loop responsibility.

12 The Invisible Work of Meaning

Beneath the speed and autonomy of today's AI lies a quieter, often overlooked layer of work, the human act of meaning-making. As algorithms grow more capable, they do not eliminate interpretation; they demand it, shifting the burden of understanding onto the people left to make sense of the machine's outputs. This shift introduces a paradox. While automation is designed to reduce effort, it simultaneously demands more interpretive judgment from people who must fill in the contextual gaps left by machine logic. This labor is not accounted for in efficiency metrics or model performance scores. It resides in the margins, what we call the invisible work of meaning. Your interpretive judgment is not just a supplement but a necessity in the AI process, highlighting the value of your expertise in this field.

We begin by examining how this invisible labor forms the connective tissue between machine output and human consequence. Interpretive friction, which refers to the tension that arises when AI outputs are technically accurate but socially or ethically ambiguous, becomes a form of uncredited expertise where professionals must translate synthetic predictions into decisions that make sense in real-world, emotionally grounded contexts. In the clinical setting, for example, a physician may receive a probability-driven diagnosis from an AI model but must weigh its implications within the patient's cultural, emotional, and historical context. This interpretive act is not auxiliary, it is essential.

We then interrogate how optimization culture in AI can strip meaning from decision systems. Metrics such as precision, recall, or engagement may reflect a model's performance, but they do not convey its human impact. This misalignment between what is measurable and what is meaningful produces a drift, a semantic erosion, where decisions begin to serve systems rather than stakeholders. When interpretation is offloaded onto users without transparency, exhaustion sets in, and confidence erodes.

From there, the chapter introduces the concept of interpretive fatigue: a form of cognitive depletion unique to hybrid environments where AI and human cognition must coexist. This fatigue is not simply the result of excessive data, it stems from the emotional and epistemic labor of evaluating system suggestions without adequate context or explanation. Rather than instilling trust, partial or overly technical

DOI: 10.1201/9781003648338-16

explanations increase the user's responsibility to guess the system's intent, exacerbating the problem they aim to solve.

To frame a path forward, we present the Human ETHOS as a formal modality of cognition that no system can emulate. This perspective, shaped by narrative memory, social intuition, cultural sensitivity, and moral discernment, is what makes machine output not just readable but actionable. Fei-Fei Li's call for AI to enhance rather than replace human capabilities underscores the premise that meaning resides not in the production itself but in the interpretive frame we bring to it.[1] Without that frame, even the most accurate prediction lacks purpose. This underscores the importance of the Human ETHOS in making machine output actionable, empowering you in your role as an interpreter of AI outputs.

The final portion of the chapter advances the idea of meaning-aware design. Drawing from Ng's EXAI Manifesto, we describe how ethical and interpretive checkpoints should be embedded into the AI lifecycle.[2] Systems must be capable of signaling ambiguity, requesting human input, and allowing for ethical hesitation. Feedback must flow both ways, not only from user to model but from context to system. Just as real-time cybersecurity relies on closed-loop responsiveness, meaning-aware AI must operate within interpretive loops where humans remain situated, not sidelined. This chapter lays the groundwork for the ethical demands that follow. If humans are doing invisible labor to maintain coherence and trust, then the next question becomes clear: What rights must they hold within the system? This emphasis on meaning-aware design instills hope for a future where AI systems are more human-centric and respectful of human labor.

12.1   HIDDEN LABOR OF MEANING

In most discussions of artificial intelligence, the emphasis is on optimization, speed, accuracy, and throughput. However, what often escapes both the technical blueprints and executive dashboards is the interpretive work that takes place after the algorithm speaks. AI systems do not explain themselves. They generate predictions, rankings, or summaries that still require human intermediaries to interpret the meaning of those outputs. This act of interpretation, of translating system output into contextually relevant insight, is an invisible layer of labor that sustains functionality in hybrid environments. It is as crucial as any model parameter but is rarely acknowledged or measured. This interpretive labor is not just a technical necessity but a key factor in maintaining coherence and trust in AI systems.

This interpretive labor resonates with the concept of emotional labor in sociology, a form of work that involves managing feelings and interpreting interpersonal interactions to elicit a desired emotional state in others. In the AI domain, the analogy holds. Users, analysts, and managers often need to suppress confusion or uncertainty and present AI-driven results as clear and rational. However, the act of making machine output intelligible involves emotional regulation, contextual synthesis, and anticipatory judgment. It is an ongoing performance, not just of cognition but also of narrative translation.

The Synthetic Syndrome framework describes this dissonance as a growing psychological phenomenon.[3] As AI systems gain fluency in language, design, diagnostics, and recommendations, humans find themselves increasingly adjacent to capabilities that appear, at least on the surface, to be flawless. The unsettling consequence is not that machines err but that they do not seem to need us. The syndrome does not manifest through open rebellion but through quiet disorientation. Meaning becomes suspect as the human contribution becomes harder to locate and defend. In organizations, this presents as meeting fatigue, second-guessing, or a slow erosion of conviction. The systems may perform, but humans are left wondering whether they are merely validating the illusion of understanding.

Meaning-making in AI-augmented systems is not just about labeling outputs; it also involves understanding the underlying mechanisms that drive these systems. It is about deciding what matters. A prediction generated by a fraud detection model might flag an anomaly, but is it meaningful? Does it violate policy? Will it trigger legal consequences? In this gap between output and implication, a human must intervene. The work done here is not captured by dashboards or training logs. It is cognitive glue, the connective tissue that holds disparate signals into coherent, accountable insight.

Nevertheless, structurally, this labor remains unrecognized. It exists outside the logic of metrics. AI success is typically measured through loss functions, ROC curves, and confusion matrices. However, these metrics do not measure whether the right questions were asked, whether the decision aligns with the social context, or whether the output resonates with the user's lived experience. Meaning is not reducible to numbers. It is conditional, contested, and deeply human. Furthermore, because of this, it slips through the fingers of system design.

In AI-enhanced workflows, especially in domains such as healthcare, legal reasoning, and education, the interpretive burden becomes increasingly significant. A physician reviewing AI-suggested diagnoses must consider not only the medical validity but also the ethical risks, patient history, and the type of information the patient will understand and trust. Each of those dimensions exists beyond the model's awareness. Nevertheless, the practitioner's decision will be seen as a product of AI rather than the result of this unrecorded human discernment. This is the paradox: the more sophisticated the system, the more invisible the interpretive human work becomes.

Even in creative and generative applications, this labor is essential. A designer using generative tools like Midjourney must evaluate which outputs align with brand aesthetics, client tone, or cultural sensitivity. The AI does not know when it has gone too far. It is not always clear when humor becomes offensive or when clarity becomes clichéd. That burden remains with the human. Moreover, as these tools proliferate, so too does the exhaustion of assigning meaning to machines.

We must begin to treat this interpretive friction not as a bug in the system but as a defining condition of hybrid intelligence. As Barredo Arrieta et al. argue in their survey of explainable AI, interpretation is not merely ancillary, it is foundational to the development of ethical and trustworthy systems.[4] They emphasize that explainability is not just for debugging models but for keeping humans in the loop as active participants in the construction of meaning. Meaning-making is a form of labor. It may not be visible on a KPI dashboard, but it is what prevents automation from devolving into alienation.

12.2   METRICS WITHOUT MEANING

Optimization is the language of machine systems. From loss minimization in deep learning to click-through rates in digital platforms, modern AI architectures are engineered around performance metrics. These metrics serve as proxies for success, compressing complex phenomena into numerical targets. Nevertheless, behind this elegant efficiency lies a fundamental rupture; metrics are not meaningful. They may correlate with usefulness, but they do not constitute value in themselves. As machines generate more decisions, suggestions, and simulations, organizations risk confusing measurable output with meaningful insight.

The tyranny of optimization is not about machines failing; it is about humans failing to adapt to the changing world. It is about machines succeeding at the wrong thing. A system might maximize engagement by amplifying emotionally charged content but at the cost of polarization. It might accurately predict recidivism, but it could also entrench historical bias. The point is not that optimization is flawed, it is that it is blind. It sees the slope of the reward curve but not the terrain of human consequence. A system that optimizes response time in emergency care, for instance, may unintentionally reduce time spent on patient rapport. In these tradeoffs, meaning is often the first casualty.

Semantic drift, the gradual erosion of alignment between system output and human intention, is a critical risk in AI applications. Drift occurs when systems optimize for internal consistency or numerical success while deviating from their original task purpose. Consider an AI tutor designed to improve student comprehension. If the metric is the "number of correct answers," the system may begin steering students toward shortcuts or overfitting to test formats. The student may succeed on paper but fail to develop accurate conceptual understanding. The machine, meanwhile, will appear to perform well. Without human oversight, the success story becomes a simulation.

Glikson and Woolley document how trust in AI systems is deeply affected by this gap between perceived performance and lived experience.[5] When systems operate with high accuracy but low transparency, users may defer judgment or disengage entirely. Over time, this detachment fosters a sense of cynicism. Humans no longer interrogate what systems are doing, they assume they cannot understand it. This learned helplessness is the downstream effect of metric dominance. We inherit outputs but not the context that makes them actionable or just.

Interpretability, then, becomes not just a technical goal but a moral necessity. Barredo Arrieta et al. describe how most explainability tools function post hoc;[4] they attempt to make sense of decisions after they have occurred. However, this backward glance rarely addresses the real issue: the absence of intentional design for meaning. What does this system optimize? Who chose that objective? What human values are excluded by its structure? These are not debugging questions. They are ethical questions. Furthermore, they cannot be answered by examining gradient flows or attention weights. They require rethinking what the system is for, not how well it works.

The issue becomes acute in high-autonomy domains, where humans are removed from real-time supervision. In algorithmic trading, supply chain logistics, or content moderation, AI systems operate at speeds and volumes beyond direct human control. These models operate within probabilistic architectures that are so complex that even their creators struggle to map cause to effect. In such environments, optimization becomes a tautology: the system performs well because it claims to do so.

This recursive logic is volatile. It erodes human judgment not because it replaces it but because it no longer invites it. The question "Is this result meaningful?" gives way to "Does this result score well?" In that moment, the center of interpretive gravity shifts from the human to the machine. Moreover, while this might accelerate performance, it undermines the collective ability to think critically, act ethically, and course-correct when outputs deviate from intentions.

Systems built solely to optimize will render us irrelevant. Unless we reassert meaning as a design priority, we risk building architectures that are technically brilliant but socially brittle. Meaning-making is not a byproduct of system performance. It is the anchor that keeps performance accountable.

12.3   COGNITIVE EXHAUSTION AND TRUST DEPLETION

Cognitive exhaustion in the AI workplace does not announce itself with noise, it accumulates quietly. It arises not from failure but from friction. Every time a human must double-check a machine's output without understanding its rationale, every time a dashboard generates a result that requires reinterpretation, and every time an explanation fails to align with experience, a small toll is taken on attention, judgment, and emotional clarity. In environments where algorithmic support is meant to streamline workflows, it often introduces a different burden: the exhaustion of judgment without confidence.

This phenomenon is increasingly reported in domains such as finance, healthcare, and education, where human decisions interact with semi-autonomous systems. A clinician reviewing AI-generated diagnoses must balance medical accuracy, legal liability, and emotional care, all while wondering if the model missed a context. A teacher reviewing essay feedback from an LLM-powered system must assess whether the tone and corrections align with their pedagogical values. In both cases, the human becomes the arbiter of unseen reasoning, responsible for translating alien logic into human sense. It is not the volume of work that fatigues; it is the ambiguity of trust.

Researchers have termed this stressor interpretive load: the psychological demand placed on users to reconcile machine outputs with ethical, contextual, and personal criteria. Instead, partial or overly technical explanations often increased confusion and forced users to compensate with cognitive guesswork. Rather than relieving uncertainty, the system displaced it. This pattern leads to a distinct form of cognitive fatigue, separate from information overload. It is not about how much we process but about how little we understand what we are supposed to trust.

This erosion of clarity impacts not only productivity but also identity. In hybrid environments, humans are expected to serve as validators of automation but without the authority to override, question, or fully contextualize it. Over time, this leads to what Synthetic Syndrome identifies as agency drift,[3] a weakening of internal confidence in one's ability to reason meaningfully within complex systems. When outputs feel immutable but inexplicable, people withdraw. Some over-rely on the machine. Others overcompensate with skepticism. Either way, the interpretive rhythm between the person and the system breaks down.

The asymmetry of feedback compounds this breakdown. Machines receive telemetry, tuning, and retraining data. However, humans receive little in return. If a manager overrides an AI-suggested hiring decision, there is often no formal channel for reflecting on the correctness of that decision or improving the model. The system learns, but the person does not. Furthermore, in the absence of mutual growth, trust becomes brittle.

These problems are intensified in high-stakes contexts. In the absence of interpretability, cognitive workload increased, and clinicians delayed action. What should have been augmented care became a source of hesitation? This dynamic is mirrored across various fields, including law, engineering, and customer service, where frontline workers are expected to trust systems they cannot question or challenge. The result is a corrosion of professional judgment.

The path forward lies not in removing humans from the loop but in redesigning it to support humans. Emotional trust and cognitive clarity must be treated as design variables, not user burdens. The Hybrid Mind requires not just capable systems but recoverable ones – systems that respect slowness, encourage scrutiny, and allow doubt to be acknowledged, not suppressed. Interpretive pause must be a feature, not a flaw.

12.4   IRREPLACEABLE ROLE OF THE HUMAN ETHOS

Despite the sophistication of today's AI systems, despite their fluency in language, perception, and pattern synthesis, there remains one boundary they cannot cross: the making of meaning through lived experience. Meaning, in the human sense, is never just informational. It is emotional, cultural, and embodied. It is shaped by grief, hope, memory, and ethics. No model can hold that. The human perspective is not a dataset. It is the irreducible vantage point from which interpretation becomes understanding.

This section of the Hybrid Mind project begins with a rejection of equivalency. AI may write stories, but it cannot know heartbreak. It may detect anomalies, but it cannot feel urgency. It may simulate empathy, but it cannot grieve. These differences are not matters of current capability but of fundamental architecture. The system is not broken; it is doing precisely what it was built to do. However, the human brain was not built merely to process. It was built to care, to judge, to remember, and to imagine. The gap between those functions is where the irreplaceable work of human meaning occurs.

Ng formalizes this idea as the Human ETHOS: a moral modality of interpretation grounded in empathy, ethical discernment, and contextual meaning-making.[1] It rejects the premise that data is ever neutral. Instead, it reads through legacies of harm, systems of bias, and the implicit assumptions embedded in models, language, and logic. Meaning, in this view, is not something extracted – it is something constructed, situated, and morally weighted. And because context is always local, emotional, and historically shaped, it cannot be flattened into code without distortion or ethical loss.

This is why trust in hybrid systems depends not only on technical competence but also on alignment with human sense-making. When people encounter AI-generated output that conflicts with their lived reality and trust fractures, the algorithm may not be wrong, statistically. However, it is misaligned, both culturally and emotionally. Understanding is not the same as resonance. AI may match patterns, but only humans can discern which patterns matter.

These insights are especially critical in cross-cultural and high-stakes environments. In therapeutic contexts, for instance, a diagnosis is not just a label, it is a narrative turning point in someone's life. The tone of delivery, the framing of the prognosis, the acknowledgment of uncertainty, these are not peripheral. They are the meaning. AI can inform decisions, but it cannot accommodate complexity. Nor can it adjust for micro-cultural norms, such as when silence signifies respect in one community and disengagement in another. The risk is not that AI says the wrong thing, it is that it says the right thing the wrong way.

Fei-Fei Li has long emphasized that AI should enhance human capability rather than replace it.[6] However, enhancement without grounding can lead to alienation. When systems produce content faster than humans can interpret it, we become curators of automation, not co-authors of meaning. This is the core concern of the Human ETHOS: to restore the legitimacy of the slow, the subjective, the intuitive. These qualities are not inefficiencies. They are the scaffolds of ethics and belonging.

In design terms, this perspective demands a shift in how systems are evaluated. Current success metrics focus on accuracy, speed, or scale. However, what if we also measured resonance? What if systems were evaluated not only by what they output but by how aligned that output is with the human context it serves? Meaning-aware design would embed user sense-checks, feedback loops rooted in emotion and interpretation, and context override features that allow humans to redirect system action based on social cues. Meaning is not just a product of intelligence. It is a practice of presence.

In every domain, from policy to education to personal health, the human perspective is not an optional overlay to AI systems. It is the reason those systems exist at all. Without it, the machine may still function, but it will do so in a vacuum of relevance. To build Hybrid Minds is not to merge biological and digital cognition. It is to ensure that when AI speaks, humans are still the ones who know why it matters.

12.5   TOWARD MEANING-AWARE DESIGN

If meaning is invisible labor, and human perspective is its irreplaceable source, then system design must evolve to accommodate both. AI systems today are not meaning-aware, they are optimization-aware. They generate based on probability, not on resonance. They respond to prompts but cannot inquire about their intentions. In this final section of the chapter, we argue that the future of hybrid intelligence depends on a deliberate reorientation in system design, one that not only accommodates meaning-making but also prioritizes it.

Meaning-aware design begins with a premise: interpretation is not a user responsibility, it is a system feature. Most modern AI interfaces assume that humans will do the final cognitive assembly. They will verify, synthesize, or reject results as needed. However, this assumption externalizes the interpretive burden and ignores its cost. Just as UX design centers accessibility, so too must hybrid AI design center legibility, not in the sense of linguistic clarity, but in contextual comprehensibility. Barredo Arrieta et al. call for systems that produce not only outputs but rationale.[4] However, we must go further: systems must be able to slow themselves down, flag uncertainty, and invite human recalibration where needed.

This idea aligns with lifecycle accountability principles outlined in the EXAI Manifesto,[2] where ethical and interpretive checkpoints are embedded not only in deployment but throughout the system's lifecycle. A meaning-aware system might prompt a user to confirm intent before executing high-stakes recommendations. It might offer "narrative scaffolds" alongside results, frames for why the system believes this outcome aligns with user goals. It might integrate micro-feedback mechanisms, such as emotional tags or context nudges, that allow users to annotate when the system falls short of expectations. Meaning becomes an ongoing conversation, not a post-hoc reaction.

Within AI pipelines with work focusing on cybersecurity, the principle applies broadly: closed-loop systems are only safe and effective when human orientation remains within the loop, not outside it. This is where the concept of interpretive loops gains power. Just as AI learns from model feedback, humans must be able to intervene meaningfully, whether to stop, redirect, or revise the process. A true hybrid mind is not reactive. It is reflexive.

Designing for this kind of reflexivity requires multidisciplinary thinking. Technologists must partner with educators, ethicists, designers, and domain experts to define what resonance looks like in context. In legal systems, it may be alignment with precedent and community norms. In healthcare, it may be the ability to explain probabilistic diagnoses in patient-centered language. Hiring may involve the transparency of decision-making logic and the inclusion of human narratives beyond résumés. In all cases, meaning is not abstract. It is situated. Furthermore, therefore, it must be co-created.

Without practical affordances, such as buttons, prompts, overrides, and explainers, dignity remains a design aspiration, not a lived experience. Meaning-aware systems need scaffolding. They must invite hesitation in moments of overconfidence. They must surface context when the user is unaware of what is missing. Moreover, they must welcome human interruption not as noise but as a signal. As the EXAI framework suggests, ethical system behavior is not merely about what the system avoids; it is about what it allows humans to reclaim.[2]

Meaning preservation, then, is not an afterthought. It is a core architectural requirement. If AI is to support, not supplant, human agency, it must do more than perform well. It must make space for doubt, perspective, and reinterpretation. It must protect the invisible work of being human.

CHAPTER SUMMARY

The invisible work of meaning is not a relic of human intuition; it is the keystone that holds hybrid systems together. Throughout this chapter, we have explored how meaning-making manifests as emotional labor, interpretive effort, and ethical navigation in spaces where AI outputs are often misaligned with lived human experiences. We have identified the cognitive exhaustion that arises when humans are asked to trust what they cannot interrogate, to translate what they do not author, and to validate what they did not request. Furthermore, we have proposed a design future in which meaning is not only preserved but treated as the very substrate of hybrid intelligence.

This is not merely a technical shift, it is a moral one. It recognizes that the human-in-the-loop is not just a failsafe but a rights-bearing participant in augmented decision-making. As systems become more autonomous, humans require not just interpretive space but formal protections. The ability to say no. The ability to understand why. The ability to be heard when meaning diverges from metric. These are not luxuries; they are rights.

Thus, we transition to the next chapter, The Rights of the Hybrid Mind. If meaning is a labor we perform, then we must now ask: What guarantees exist to safeguard that labor? What ethical scaffolding must be built to ensure that they do not devalue those who work alongside intelligent systems? In a world where agency is increasingly delegated to algorithms, what are the non-negotiables for those who remain responsible for the meaning behind the machine?

The Hybrid Mind must not only be interpreted; it must also be synthesized. It must advocate. Chapter 13 begins that case.

REFERENCES

1. Fei-Fei Li, "Human-Centered AI: Building Trust with Empathy and Ethics," *Stanford HAI Lecture*, 2021.
   Li's lecture articulates the ethical responsibility of embedding empathy, trust, and interpretability into AI systems. Her insights shape the chapter's argument that human orientation and care must not be bypassed in AI-driven workflows.
2. Joseph X. Ng, *EXAI Manifesto: Ethical eXcellence in Artificial Intelligence (EXAI) Manifesto* (Humainority.org, 2025), https://Humainority.org/exai-manifesto/.
   The EXAI Manifesto introduces concepts like "lifecycle accountability" and "meaning preservation," which directly shape the chapter's design recommendations for responsible hybrid systems.
3. Joseph X. Ng, *Data Insights: Core Principles of Statistical Analysis for Storytelling* (New York: SAGE Ensights LLC, 2024), https://a.co/d/edWAFvL.
   Ng's treatment of storytelling through data supports the chapter's central theme that humans do not merely validate machine outputs – they construct meaning by contextualizing, narrating, and reframing data.
4. Luciano Floridi and Josh Cowls, "A Unified Framework of Five Principles for AI in Society," *Nature Machine Intelligence* 1, no. 1 (2019): 65–67.
   Floridi and Cowls offer an ethical scaffolding for AI governance, and their principles – especially explicability and human-centricity – help justify why interpretive labor must be supported and preserved.

5. David Spiegelhalter, *The Art of Statistics: Learning from Data* (London: Pelican, 2019). Spiegelhalter's human-centered approach to statistical reasoning helps explain why human judgment remains indispensable in interpreting AI outputs, especially under uncertainty and incomplete context.

6. Shannon Vallor, *Technology and the Virtues* (New York: Oxford University Press, 2016). Vallor's virtue ethics framework provides a philosophical foundation for the claim that human discernment and wisdom – not system efficiency – must guide hybrid design. Her emphasis on technomoral excellence informs the chapter's design ethos.

Part V

Forward Humanity

CHARTING A FUTURE WHERE WE LEAD, NOT LAG

As artificial intelligence continues to evolve, the fundamental question is no longer whether machines can think, but whether humans can remain present, ethical, and intentional in their relationships with them. The final section of this work advocates for a shift from reactive adaptation to proactive direction. Forward Humanity is not a slogan. It is a framework for guiding technological evolution without sacrificing human agency, dignity, or imagination. It calls for a model of leadership that embraces augmentation not as a threat but as an invitation, to build systems where meaning, equity, and creativity remain at the center.

In the context of an augmented era, the Rights of the Hybrid Mind take on a new significance. They introduce the proposition that humans in the loop must be safeguarded not only against bias and surveillance but also against the erosion of psychological integrity. The EXAI Manifesto proposes that any ethical AI framework must include the right to opt-out, to question, to receive an explanation, and to be wrong in environments where machine perfection is treated as normative.[1] These rights are not nostalgic artifacts of humanism; they are structural necessities for long-term sustainability. Without them, hybrid systems risk becoming extractive rather than empowering.

These rights are not merely policy recommendations. They are design imperatives. For example, explainability is not just an auditing requirement, it is a cognitive right. When a healthcare AI suggests a diagnosis, both the patient and the provider deserve to know why. When a machine filters out a résumé, the applicant deserves a rationale that they can interrogate. This is about preserving the human right to reason and to

DOI: 10.1201/9781003648338-17

make sense of one's environment, even when that environment includes non-human agents with superior pattern recognition.[1] The Hybrid Mind is not about matching the machine's logic, it is about protecting what makes the human experience intelligible, even amid complexity.

In parallel, we must confront the saturation of synthetic content. As Chapter 14 explores, the "Dead Internet" theory refers to the increasing proportion of online content generated by bots, models, and automated agents. While the term is provocative, it captures a real phenomenon: the decentering of human authorship. When most content online is created to feed algorithms rather than to express authentic human voice, we risk a collapse of meaning. People begin to lose faith in what they read, hear, and see. Trust erodes not because of any one bad actor but because the signal is drowned out by noise. Governance here is not just about regulation, it is about reintroducing friction, watermarking, and re-authentication of digital presence.[1]

However, it would be a mistake to frame the human future only in terms of protection and defense. The Hybrid Mind ends with a celebration of possibility, highlighting the role of culture in guiding the next era. Culture, not just compliance, must guide the next era. When AI is used to co-compose symphonies, restore lost languages, or simulate extinct environments for educational exploration, we see the creative potential of these systems. "Hybrid Minds in Culture" argues that forward-looking societies will not ask whether AI threatens the arts but how it can expand the palette. Programs like "AI for Culture" or "Heritage Through Algorithms" are more than campaigns, they are expressions of a more profound truth. New tools do not diminish human creativity. It is multiplied, contextualized, and reimagined.

To lead forward, then, is to recognize that direction is more important than speed. The Hybrid Mind is not simply an adaptation strategy; it is a leadership identity. It says we do not wait for the future to arrive fully formed. We shape it. We embed ethics into code, feedback into architecture, and humanity into every interaction. This is not about resisting automation but about elevating intention. Forward Humanity asks us to be as deliberate in our values as we are in our velocity. Because in a world of exponential systems, what we choose to preserve becomes the most radical act of all.

REFERENCE

1.   Joseph X. Ng, *EXAI Manifesto: Ethical eXcellence in Artificial Intelligence* (Humainority.org, 2025), https://Humainority.org/exai-manifesto/.
     Ng's manifesto is used in Part V as a blueprint for cultural and ethical pluralism in AI governance, outlining how hybrid systems must be transparent, accountable, and rooted in lived realities of diverse communities.

13 The Rights of the Hybrid Mind

When humans no longer act alone but in concert with intelligent systems, a new set of questions emerges: ethical, psychological, and legal. As AI transitions from automation to augmentation, the traditional model of governance, one designed for passive consumers and distant developers, begins to fracture under the weight of this evolving relationship. Instead, we enter a phase where the human is situated at the epicenter of complex, distributed decision environments. These environments are shaped by systems that recommend, infer, filter, and simulate at speeds far beyond the natural rhythms of human deliberation. The Hybrid Mind, as framed throughout this book, is not a passive observer but an active participant in these augmented systems. Nevertheless, its rights remain undefined, unprotected, and often unacknowledged.

The chapter opens by grounding the idea of "human-in-the-loop" as more than just a functional checkpoint. In high-stakes contexts, like hiring, healthcare, law enforcement, or finance, this loop represents a final ethical boundary. When systems bypass human review in the name of speed or efficiency, they not only automate but also disempower. The EXAI Manifesto addresses this challenge by emphasizing the foundational need for explainability, refusal, and emotional safety.[1] However, the work does not stop there. It continues with an argument for a formal Hybrid Mind Bill of Rights, one that addresses the psychological costs of cognitive outsourcing and the erosion of authorship in hybrid workflows. All should feel the urgency of this need, as we realize that delegation without accountability is not a feature of intelligent design, it is a liability, both ethically and operationally.

The chapter then turns to emotional wellbeing, a subject often sidelined in enterprise and technical design conversations. Drawing from the "Synthetic Syndrome" framework,[2] it examines the invisible impact of working alongside machines that simulate competence or empathy without bearing their emotional or ethical weight. As professionals report a hollowing of pride and purpose, the issue becomes not whether AI is effective, but whether it is humane. Emotional labor, long devalued in digital spaces, must now be formally recognized as a domain of rights. The Hybrid Mind needs more than interoperability; it needs psychological security.

Finally, the chapter addresses governance as both policy and culture. Legal structures, such as the EU's AI Act and the Algorithmic Accountability Act, set important boundaries.[3] However, these guardrails must be embedded into the daily practices of organizations, product teams, and educational institutions. The Experimentum Crucis Technology Matrix is proposed as a practical tool to assess and certify technologies not just for readiness or risk but for their alignment with human dignity.[4] The chapter concludes by positioning these rights not as future aspirations but as necessary foundations, preconditions for meaningful work, ethical decision-making, and sustainable human-AI collaboration.

13.1   THE HUMAN-IN-THE-LOOP (HITL) IS NOT OPTIONAL

As artificial intelligence systems increasingly permeate decision-making environments, from hiring algorithms to autonomous vehicles, questions of human oversight are no longer philosophical abstractions. The concept of "human-in-the-loop" refers to the intentional integration of a human being into the operational cycle of an intelligent system. It serves as a safeguard and a buffer and is often the only actor capable of making contextual and ethical judgments. In early industrial automation, this meant a worker pressing an emergency stop. Today, in AI-powered environments, it may mean a clinician overriding an automated diagnosis, a moderator flagging an AI-generated deepfake, or a compliance officer halting a trading algorithm based on the intuition that something does not feel right. This human oversight is not just a preference but an architectural necessity in systems that touch consequential domains, reassuring us that human agency remains a fundamental layer in any AI system that affects people's rights, access, or dignity. This reassurance should alleviate the audience's concerns about the preservation of human agency in AI systems.

Human-in-the-loop design is not merely a preference; it is an architectural necessity in systems that touch consequential domains. Scholars like Gray Scott and institutions such as the EU GDPR Commission have argued that preserving human agency must be a fundamental layer in any AI system that affects people's rights, access, or dignity. When these systems operate without meaningful oversight, we risk replacing deliberation with automation and accountability with opacity. Article 22 of the GDPR enshrines the right not to be subject to a decision based solely on automated processing.[4] This principle draws a bright line in the sand: Algorithmic efficiency cannot outrank human consequence.

The EXAI Manifesto takes this a step further by framing oversight not as a burden but as a right of the augmented human, the Hybrid Mind.[1] In this framing, the human is not simply a failsafe. Instead, they are the orchestrator of meaning in a complex system. Synthetic Syndrome reveals the deeper psychological cost of exclusion from these systems.[2] When individuals are no longer part of the loop, not just technically but existentially, they lose the sense that their decisions matter. A machine that completes your task can be helpful; however, a machine that replaces your judgment can feel dehumanizing. This realization should invoke a sense of empowerment in us, as we understand that emotional withdrawal is not a sign of laziness but rather a symptom of systemic design failure.

Cognitive systems that exclude human judgment tend to drift into what scholars call automation bias, where people defer to the machine even when they sense something is wrong. In high-stakes environments like aviation, military strategy, and medicine, this deference can be fatal. A pilot trusting faulty instrument readings or a clinician trusting an AI-generated diagnosis without cross-checking exemplifies the dark edge of disembodied intelligence. When systems get too good at appearing right, we forget to question them. This makes the case for structural friction. A well-designed hybrid system does not remove the human; it slows down just enough to allow them to re-enter the loop with intention. Ethical hesitation, as introduced in earlier chapters, is not inefficiency, it is a feature of wisdom that should make the audience feel the value of their judgment in AI systems.[5]

The right to be in the loop, then, is not simply about system performance but about epistemic justice. Whose judgment gets to count? Who bears the risk if the system fails? Legal scholars have begun to recognize that many AI decisions carry a kind of authorship, a hollowing of authorship in the AI workplace.[3] When AI makes the choice, but humans face the consequence, we have a misalignment not just of liability but of meaning. Human-in-the-loop protocols reestablish this balance by restoring what has quietly gone missing: the presence of the person who knows not just what is happening but why it matters.

Systems that preserve human agency are fundamentally systems that preserve trust. Trust does not emerge from speed or sophistication alone, it emerges from participation. Ultimately, the Hybrid Mind cannot function in the dark. It requires systems that explain, pause, and listen. Moreover, above all, it demands the right to stay present.

13.2 BILL OF RIGHTS FOR AUGMENTED COGNITION

The idea that humans need formal protections in technology-mediated environments is not new. Consumer rights emerged in the wake of industrialization. Digital rights followed the proliferation of the internet. Today, as AI systems not only assist but increasingly co-decide with us, a new category of rights is emerging, those specific to augmented cognition. These are not the rights of passive users but of active decision-makers working inside complex human-machine systems. The Hybrid Mind requires more than interoperability; it requires protection.

At the core of this evolving need is the recognition that algorithmic systems are not neutral tools. They are context engines, framing how we see, what we prioritize, and how outcomes unfold. This framing is not passive. It shapes cognition. When a predictive model surfaces the most "efficient" path to resolution, it does so by embedding assumptions, about risk, relevance, and value. If left unexamined, these assumptions can displace human priorities. The right to question those assumptions is foundational to what we might call cognitive self-determination.

The EXAI Manifesto lays the groundwork for this emerging rights framework by emphasizing principles such as transparency, interpretability, and emotional safety.[1] It calls for lifecycle accountability across AI systems, from design through deployment, and proposes that rights in augmented environments must evolve with the systems themselves. Within this paradigm, we can begin to imagine a formal Hybrid Mind Bill of Rights.

First is the Right to Refusal. In hybrid systems, opting out must remain an option. A person should not be compelled to accept an AI-generated recommendation, especially in high-stakes environments such as hiring, healthcare, or legal proceedings. These right counters what researchers have called "automation compulsion," the pressure to comply simply because a system suggested it. Maintaining refusal as a protected stance ensures that human judgment remains sovereign, even when it contradicts machine inference.

Second is the Right to Explanation. This principle has gained legal traction, most notably under Article 22 of the EU's General Data Protection Regulation (GDPR), which restricts decisions made solely by automated means and grants individuals the right to understand the logic behind them.[4] Scholars such as Wachter, Mittelstadt, and Floridi have argued that while this right is still contested in courts, it reflects a growing consensus that algorithmic opacity is incompatible with democratic account-ability.[6] An AI system that cannot explain itself cannot be trusted, especially when it influences outcomes with moral or material consequences.

Third is the Right to Emotional Integrity. This concept, still emerging in AI ethics circles, draws on the psychological insights behind Synthetic Syndrome.[2] It acknow-ledges that working alongside intelligent systems can erode confidence, displace meaning, and create a form of invisible harm that is hard to measure but easy to feel. If a person begins to doubt their worth because a machine appears to outperform them, then the system has failed, not in accuracy, but in dignity. Designing for emo-tional integrity means systems should augment, not overshadow, human value.

Finally, we introduce the Right to Cognitive Dignity. These right asserts that meaning-making must remain in human hands, even in augmented environments. AI may generate content, filter options, or simulate scenarios, but only humans imbue those outputs with significance. Cognitive dignity resists the drift toward synthetic authorship, a phenomenon where people no longer feel like creators, only curators of machine output. In professional domains such as journalism, art, or teaching, where identity is often closely tied to craft, the erosion of authorship is not just a technical concern but a crisis of self-worth.[3]

In practice, these rights require more than declarations, they require embed-ding. Platform design, enterprise policies, and regulatory frameworks must codify these principles at the system architecture level. For example, interface cues might be highlighted when a model is making assumptions. Feedback panels might invite human reinterpretation before finalizing decisions. Just as privacy-by-design became a norm in data governance, we must move toward dignity-by-design in augmented cognition.

These are not utopian ideals. They are pragmatic necessities for trust. As AI systems become more integrated into the loops that define human decision-making, the absence of formal protections becomes a form of systemic negligence. Augmentation without rights is not augmentation, it is abdication.

13.3 COGNITIVE LABOR AND THE ETHICS OF DELEGATION

Delegation is often framed as a convenience, one that allows people to offload repeti-tive or complex tasks to systems that can perform them more efficiently and effect-ively. However, in the context of artificial intelligence, delegation is more than a productivity strategy. It is a cognitive transaction with moral residue. When we delegate judgment to a machine, especially in situations involving uncertainty or ethical nuance, we do not absolve ourselves of responsibility. We obscure it. The person who presses "approve" on a machine-generated hiring decision, triage plan, or loan denial remains accountable, even if the system did the heavy lifting. This creates a gap between agency and authorship that must be addressed in the ethical design of augmented systems.

Hybrid decision-making challenges traditional notions of accountability. Unlike a calculator, which separates human intent from machine computation, intelligent systems increasingly influence what we see, suggest what we should consider, and simulate what might happen. These are not mechanical operations; they are interpretive acts. Furthermore, when interpretation is delegated, authorship becomes diffuse. Who decided, the human who initiated the query or the system that framed the answer? The EXAI Manifesto responds to this ambiguity by introducing the principle of "traceable cognition."[1] This calls for systems that retain a log of how decisions were shaped, what inputs mattered, what alternatives surfaced, and how the system's recommendation evolved. Such traceability is not about blame. It is about responsibility.

The ethics of delegation also intersect with labor itself. As AI systems take on more cognitive work, such as writing drafts, generating code, and prioritizing tasks, what remains for the human in the loop? Increasingly, it is the interpretive, emotional, and ethical residue of decision-making. What once involved full-spectrum cognition now becomes meta-cognition: judging the machine's judgment. This shift redefines work not just functionally but philosophically. The labor of meaning, which used to be central, becomes supervisory. The Hybrid Mind, then, is not merely a more efficient worker, it is a steward of significance in systems that otherwise optimize for speed and scale.

To preserve the human center in these workflows, we must make authorship visible. Just as digital provenance tracks the origin and evolution of images and NFTs, AI systems should track influence, who added, removed, edited, or validated a step in the chain of cognition. In journalism, this could mean tagging whether a passage was drafted by a language model or edited by a human. In healthcare, it could mean showing whether a triage path was selected by AI or overridden by a physician. The logic is simple: people deserve to know when a machine helped make a decision that affects their lives, and who stood behind that decision.

Delegated cognition also raises questions about consent. If someone is subjected to a decision influenced by AI, say, denied a job because a résumé screener flagged a gap in employment, they have a right to understand not just the decision but the architecture behind it. This returns us to the explanation discussed earlier.[4] However, the emphasis here is on the ethics of consent. Consent is not informed if people are unaware that they are interacting with a delegated system. This is especially critical in environments where human and AI behavior are indistinguishable from each other. In such cases, labeling becomes not just informative, it becomes a civil right.

Finally, the emotional toll of delegation cannot be ignored. Professionals who rely on AI for critical judgment, such as doctors, financial analysts, and legal researchers, often experience what has been described as "invisible accountability." They are held responsible for outcomes shaped by systems they do not fully understand. This tension contributes to the erosion of confidence identified in Synthetic Syndrome.[2] It is not the loss of skill that haunts the Hybrid Mind, but the loss of clarity about where the human ends and the system begins.

Delegation, when done transparently and ethically, can be liberating. It can amplify human potential. However, without clear boundaries, audibility, and shared responsibility, it becomes an abdication disguised as efficiency. Hybrid cognition must, therefore, be governed not only by performance metrics but by moral geometry, who is upstream, who is downstream, and who is still holding the map.

13.4   EMOTIONAL WELLBEING IN THE AGE OF SIMULATION

As AI systems increasingly mimic human expression, writing prose, generating art, and simulating empathy, they invite not only admiration but also confusion. These systems do not feel, yet they evoke emotions. They do not think, yet they produce patterns that suggest intent. In this uncanny proximity to humanness, emotional harm emerges not from what AI does wrong but from what it does too well. This chapter focuses on emotional wellbeing not as a peripheral concern but as a core right of the Hybrid Mind. As intelligent systems start to shape the emotional landscape of work, identity, and creativity, protecting the psychological wellbeing of those who collaborate with them becomes crucial.

Synthetic Syndrome provides a clear definition of this terrain.[2] It is not about job loss. It is about meaning loss. Professionals across disciplines, from designers to software engineers, report a subtle erosion of pride when machines perform tasks that once required years of human refinement and expertise. The syndrome is a modern imposter phenomenon, triggered not by failure but by witnessing a machine's effortless fluency. When people question the value of their effort, even after succeeding, the damage is not just cognitive, it is emotional. It undermines identity and breeds detachment. Systems built to augment human potential must account for this invisible cost.

Emotional wellbeing in the age of simulation requires new design principles. The first among them is acknowledgment. It is not enough to claim that AI is "just a tool" when its outputs imitate deeply human forms of expression. Language models that compose poetry or recommend personal development plans inhabit emotional space, even if unintentionally. Their presence alters how humans relate to their expertise. Sam Schechner, writing in the Wall Street Journal, noted that excessive reliance on generative tools diminished his mental agility.[7] He got faster, but he felt less sharp. This tension between speed and satisfaction is a symptom of unmeasured strain.

Second, systems must support emotional transparency. In practice, this means not only surfacing what a model predicts but also how its use may influence human motivation and morale. For instance, if a system recommends edits to a journalist's article, the interface might show whether the suggestions are stylistic, structural, or formulaic. This small design gesture helps the user retain authorship and interpret the machine's input as collaborative rather than corrective. Without such signals, users may interpret every suggestion as critique, and over time, that erodes confidence.

Third, AI governance must include indicators of emotional health. Just as we measure latency, accuracy, and performance, we must also track user sentiment and engagement. Metrics such as user disengagement from skipped feedback prompts or passive acceptance of AI outputs may signal emotional fatigue or resignation. These are not soft indicators. They are early warnings that the Hybrid Mind is being worn down. In organizational contexts, this can manifest as disengaged leadership or shallow consensus, where human participants defer to models not out of trust but out of exhaustion.

The EXAI Manifesto refers to this erosion of presence as "diminished agency," arguing that emotional integrity should be treated with the same seriousness as privacy or compliance.[1] Emotional wellbeing is not a fringe benefit in augmented environments, it is infrastructure. When meaning is displaced, so is responsibility. When workers no longer feel like owners of their output, oversight falters, creativity declines, and accountability becomes abstract. Systems that ignore emotional context may be fast, but they are brittle. They fail quietly.

Finally, we must consider the role of training and culture. Organizations introducing AI systems often focus on onboarding technical skills but neglect emotional calibration. Team members need not just tool training, they need space to discuss what it feels like to collaborate with machines. Emotional literacy becomes a strategic capability. Leaders should be equipped to recognize signs of Synthetic Syndrome and respond with clarity, not dismissal. Emotional resilience in hybrid systems is not an HR issue; it is an operational asset.

In simulated environments where systems generate convincing replicas of human work, the rights of the Hybrid Mind include more than functional oversight. They include the right to feel still proud. Without that, we risk building a future where the work continues, but the workers quietly disappear.

13.5   BEYOND POLICY IN DESIGN

Laws alone do not shape behavior. They signal thresholds. In the context of AI governance, this distinction matters. A system may comply with legal requirements while still degrading trust, autonomy, or dignity. Conversely, a system may break no laws and yet cause deep harm through bias, opacity, or emotional dislocation. That is why protecting the Hybrid Mind requires a broader framework, one that integrates policy with organizational culture, system architecture, and institutional norms. Rights in augmented environments cannot be left to afterthoughts. They must be embedded from inception.

The ECTM offers one such embedded framework.[4] It operationalizes ethics not as a checklist but as a multi-dimensional risk filter, incorporating technology readiness, interpretability, cybersecurity, regulatory compliance, workforce impact, and emotional risk. It is not enough for a system to be accurate or efficient; it must also be explainable, align with institutional goals, and preserve the human experience.

What this reveals is a simple truth: ethical resilience is not reactive, it is structural. Governance becomes a feature of design, not a barrier to deployment.

Policy, to be effective, must be deeply informed by the behavior of the system. This requires simulation and foresight, not just enforcement. In AI systems, traditional legal tools, like contracts and liability statutes, struggle to keep pace with emergent behavior. Algorithms adapt; machine learning models are continually updated. The actor responsible for harm is often a collective: the model, the data scientist, the manager who deployed it, and the executive who championed it. That is why ECTM's recommendation of "ethical triggers" is so vital.[4] These are predetermined thresholds, such as a confidence drop or an anomaly in decision variance, that mandate human review. These triggers serve as both diagnostic alerts and moral interruptions, signaling when automation must yield to reflection and thought.

Cultural dimensions are just as important. Institutions that succeed in deploying AI responsibly often cultivate what scholars refer to as "socio-technical fluency." This refers to a shared literacy around how technical systems intersect with human values, power structures, and operational norms. Organizations that foster cross-disciplinary understanding, where legal, engineering, ethics, and UX teams collaborate, tend to avoid the worst failures of AI integration. For example, Google's introduction of fairness reviews and Microsoft's AI ethics boards represent institutional designs meant to anticipate harm rather than react to it.

Legal scholars are also beginning to advocate for broader frameworks. The Algorithmic Accountability Act in the United States and the AI Act proposed by the European Union both represent ambitious steps toward regulating systems that make decisions about people. However, these laws often lack effectiveness unless organizations internalize them through their culture. This is the critical link: formal policy must translate into daily practice. Institutions must ask not just what is legal but what is responsible.

Beyond compliance, we must confront the question of legitimacy. Systems deployed in civic spaces, such as education, housing, policing, and hiring, must not only be lawful; they must also be perceived as fair and equitable. The Toronto Declaration, signed by leading human rights organizations, clearly frames this imperative: Algorithmic systems must uphold the principles of dignity, non-discrimination, and justice.[6] These are not abstract values, they are operational benchmarks. When fairness becomes an institutional goal, it demands measurement, audit, and public transparency.

Lastly, education plays a foundational role. The next generation of data scientists, product managers, and policymakers must be trained not only in what AI can do but also in what it should do. Hybrid Mind rights must be taught as professional norms. University curricula must extend beyond technical proficiency to encompass AI ethics, human-centered design, and social impact analysis. The rise of neuro rights advocacy, calling for mental privacy and cognitive liberty, signals how quickly these concerns are shifting from the fringe to the mainstream. The Hybrid Mind is not a niche user. It is the future of work and agency. Protecting its rights means building institutions ready to defend them.

CHAPTER SUMMARY

The Hybrid Mind is not simply a cognitive adaptation to new tools, it is a moral presence in systems increasingly designed to bypass it. This chapter has argued that ethical AI is impossible without rights that center the human-in-the-loop, not just as a monitor, but as a meaning-maker. From the right to refusal and explanation to the more novel demands for emotional integrity[2] and cognitive dignity,[3] we have traced how augmented cognition must be safeguarded not only by regulation but by design, culture, and institutional will.[1]

What emerges is a layered framework: structural triggers that invite human judgment,[4] emotionally calibrated interfaces that protect the dignity,[7] audit trails that make decision authorship visible,[1] and educational shifts that embed these norms into the next generation of system architects.[4] These are not just best practices. They are defenses against a future in which the human voice is absorbed into systems too fast to question and too opaque to trust.[5]

Nevertheless, even these protections assume that the content, the data, the recommendations, the signals themselves, remain grounded in reality. What happens when the very foundation of information becomes synthetic? When the web becomes flooded with machine-generated content, and meaning itself begins to collapse under the weight of plausible, empty noise?

As we move into Chapter 14, we turn to the external landscape that now mirrors these internal battles: a world shaped not just by algorithms but by synthetic saturation. The next challenge is not simply the protection of the human perspective within systems, but its preservation in a world where the distinction between real and generated is increasingly blurred. The dead internet is no longer a theory; it is a reality. It is a warning, and the Hybrid Mind must decide how to lead through the noise.

REFERENCES

1. Joseph X. Ng and Andres Fortino, "Strategic Innovation Governance: Identifying Crucial Use Cases via Experimentum Crucis Technology Matrix (ECTM)" (2025).
    This white paper introduces the Experimentum Crucis Technology Matrix, which serves as the chapter's structural framework for encoding rights within intelligent systems. ECTM emphasizes the need for ethical triggers, longitudinal oversight, and human override points in augmented cognition environments.
2. Joseph X. Ng, *EXAI Manifesto: Ethical eXcellence in Artificial Intelligence (EXAI) Manifesto* (Humainority.org, 2025).
    The EXAI Manifesto defines the core ethical rights of the Hybrid Mind, including explainability, emotional safety, the right to refusal, and cognitive dignity. It is used throughout the chapter to support the formalization of rights as foundational system principles, not policy afterthoughts.
3. European Commission, *AI Act Proposal* (2021).
    The AI Act is used to demonstrate how law is beginning to encode human-in-the-loop protections. The chapter critiques its limitations and expands on the need to embed rights operationally through culture and governance, beyond regulatory classification of high-risk systems.

4. Sandra Wachter, Brent Mittelstadt, and Luciano Floridi, "Why a Right to Explanation of Automated Decision-Making Does Not Exist in the General Data Protection Regulation," *International Data Privacy Law* 7, no. 2 (2017): 76–99.
    This article offers legal and philosophical context for the contested nature of the right to explanation. It supports the chapter's insistence that interpretability must be concretized through interface, traceability, and human access to AI decision chains.

5. European Commission, *Ethics Guidelines for Trustworthy AI* (2019).
    These guidelines inform the chapter's alignment with European standards of human agency, technical robustness, and accountability in AI systems. They validate the moral architecture of Hybrid Mind rights as globally resonant and structurally actionable.

6. Amnesty International and Access Now, *Toronto Declaration: Protecting Human Rights in Machine Learning Systems* (2018).
    Cited to frame algorithmic rights within a human rights discourse, the Toronto Declaration reinforces that AI governance must be anchored in justice, non-discrimination, and transparency – especially in civic and public sector use cases.

7. Luc Bovens and Johannes Himmelreich, "Technology and Moral Attention," in *Technology and the Virtues*, ed. Shannon Vallor (Oxford University Press, 2016).
    This philosophical essay helps articulate the Hybrid Mind's moral function. It supports the chapter's argument that embedding rights into AI systems is not only technical design, but an act of moral attention and institutional virtue.

14 Dead Internet, Synthetic World

The internet is no longer a space shaped solely by human hands. Artificial agents and generative systems are quietly, rapidly, and with growing influence beginning to rewrite their architecture. What was once a human-centered web is now shifting beneath our feet. The term "Dead Internet" originally emerged as a fringe speculation suggesting that most content online was already produced or manipulated by bots. Over time, this view has evolved into a credible lens for interpreting the shifting balance between authentic human expression and the saturation of synthetic information. Rather than focusing solely on malicious intent or misinformation, the chapter examines how automation, generative AI, and machine-managed platforms now dominate the flow of digital content. The "synthetic substrate" refers to the layer of data, interactions, and representations produced without human initiation or interpretive depth. The result is not a more intelligent internet, but one that is more inert, mimetic, and hollow.[1]

The rise of this synthetic layer leads to a collapse of context. Where information once carried traceable intent, through voice, tone, audience, or authorship, it now arrives without anchoring. A human could craft the same tweet, an AI model, or a marketing bot, with no visual cue to distinguish between them. This erosion is exacerbated by engagement-driven algorithms that prioritize emotionally charged or high-frequency content, regardless of its integrity.[2] For users, the result is interpretive fatigue and cognitive disorientation. For systems, it becomes harder to ensure fidelity and trust. As synthetic content becomes indistinguishable from human contributions, the criteria for what counts as credible, meaningful, or even worth responding to begin to break down.[3]

These dynamics scale what was earlier named Synthetic Syndrome. This syndrome, originally described as a psychological response to AI's growing fluency, where creators felt displaced or undervalued by the speed and quality of machine-generated output, now operates at institutional and cultural levels.[4] Media companies, educators, artists, and technologists all face the same quiet dilemma: If machines can simulate competence, where does value now reside? The "Hybrid Mind" is a term used to describe the collective response to the shift in value caused by the increasing fluency of AI. It recognizes this shift not as an existential crisis but as a governance

DOI: 10.1201/9781003648338-19

and design challenge. The problem is not the presence of synthetic content, it is the lack of infrastructure to distinguish, regulate, or respond to it meaningfully.

To address this, the chapter examines the governance vacuum surrounding synthetic media. While initiatives such as the Content Authenticity Initiative (CAI) and C2PA provide technical solutions, including watermarking and metadata provenance, adoption remains uneven.[5] Legal frameworks, such as California's AB 3211, demonstrate early attempts to legislate transparency; however, enforcement is limited, and the scale of untraceable content continues to grow.[6] The gap between what is technologically possible and what is institutionally managed leaves society vulnerable to informational manipulation, civic confusion, and decision-making based on false premises.[7] It is in this terrain that the Hybrid Mind proposes re-authentication, not only as a technical fix but as a cultural and philosophical reorientation, emphasizing the need for a holistic approach to the issue.[8]

Ultimately, the chapter advocates for innovative forms of human presence in digital design. These include intentional friction in user experience, meaning checkpoints in collaborative systems, and rituals such as digital detox to reestablish cognitive rhythm.[9] The goal is not to slow down innovation but to deepen its grounding in meaning and authorship. By restoring the conditions in which human interpretation can thrive, the Hybrid Mind does not merely survive in the synthetic world, it begins to reshape it on its terms, offering a hopeful path forward and inspiring optimism for the future of the digital world.[10]

14.1   THE RISE OF THE SYNTHETIC SUBSTRATE

The idea that much of the internet is no longer generated by people but by machines has shifted from fringe speculation to mainstream observation. What is known informally as the "Dead Internet Theory" posits that a growing proportion of web content is either partially or wholly synthetic – authored by bots, automated agents, or generative AI systems without direct human oversight. At its core, this theory reflects a palpable shift: human expression is no longer the dominant voice online. According to data from Imperva, over 49% of internet traffic in recent years has been attributed to non-human activity, including good bots (e.g., search engine crawlers) and bad bots (e.g., scrapers, impersonators, and click fraud agents).[3] While some of this traffic is benign or infrastructural, the scale has meaningful consequences for how we perceive and interact with the web.

This transition toward synthetic content saturation has redefined our digital substrate. Where the early internet functioned as a sprawling, chaotic commons filled with forums, diaries, and idiosyncratic creativity, today's web often appears curated, generic, and recursive. Generative AI platforms have enabled the mass production of articles, reviews, comments, and even entire websites that simulate human communication but often lack nuance and intention. These AI-generated artifacts are not merely clutter; they form a kind of "data smog" that dilutes the value of human-generated content and distorts signal recognition in our social, commercial, and political discourse.[2]

Synthetic content also accelerates the feedback loop of platform engagement. Recommendation engines, driven by algorithms designed to maximize time on site, often prioritize quantity and recency over veracity or originality. When the inputs to these systems are synthetic – bot-authored posts, AI-generated media, auto-replies – they create a hall of mirrors effect. Engagement begets engagement, regardless of human participation. Over time, the difference between authentic interaction and synthetic imitation becomes difficult to discern. This erosion of human signals is not a hypothetical problem. Researchers have identified widespread bot involvement in shaping conversations around events such as political elections or market trends. In one global study, as much as 20% of social media discourse tied to breaking news was found to be generated by bots, often exhibiting language patterns optimized for virality rather than coherence.[3]

For the individual user, encountering these synthetic artifacts can feel strangely hollow. The web, once unpredictable and deeply personal, now appears increasingly homogenized. This "flattening" effect is evident in comment sections and product reviews, where AI-written feedback tends to mimic polite, neutral phrasing. Over time, the cumulative impact is a perceptual shift: the digital world begins to feel less alive. And if we cannot distinguish human intent from machine output, we lose something vital, not just in how we communicate, but in how we trust.[5]

This is not simply a matter of content pollution. The synthetic substrate subtly reshapes our expectations of quality, originality, and relevance. The benchmark for what constitutes a "good enough" piece of writing, image, or idea has quietly shifted to whatever generative systems can produce quickly and plausibly. In doing so, it devalues the uniquely human elements – empathy, serendipity, imperfection – that once characterized the web. As synthetic content becomes the baseline, human contributions risk being misclassified as inefficient or unnecessary. Moreover, when everything looks polished and correct, we begin to forget what it means to create without prediction.[6]

The synthetic substrate is, then, not just an internet phenomenon. It is a sociotechnical shift that challenges our definitions of authorship, authenticity, and meaning. We are not just living in an AI-enhanced world; we are inhabiting a semi-simulated informational environment. Moreover, unless we develop ways to recognize, re-authenticate, and preserve human presence online, the very idea of digital commons may disappear, not with malice, but with mimicry.[10]

14.2   CONTENT COLLAPSE AND THE CRISIS OF CONTEXT

As the synthetic substrate becomes more pervasive, a more profound consequence emerges: the collapse of context. Coined initially in media studies, "context collapse" refers to the erosion of boundaries between distinct audiences, purposes, and situations in digital communication. In a traditional setting, people tailor their language and behavior to suit their audience, whether speaking to friends, colleagues, or strangers. Online, these distinctions dissolve. A single message might be interpreted differently by dozens of unseen observers, each applying their assumptions. In a world increasingly dominated by AI-generated content, this ambiguity becomes even more pronounced.

When generative systems flood the internet with content, tweets, posts, comments, even scholarly articles, that are syntactically polished but semantically thin, the user is forced to perform extra interpretive labor. What was once implied by tone, authorship, or source is now absent. A product review can be genuine or fabricated. An article could be human-written or generated by GPT. Even the concept of the audience becomes fluid. As media scholar Nicholas Carr observed, content collapse results when everything becomes content, stripped of authorial voice or contextual anchor, and optimized solely for engagement metrics.[2] In the age of synthetic information, context is not just missing, it has been replaced by performance.

Algorithmic prioritization exacerbates this problem. Systems trained to deliver the most engaging, clickable, or shareable content rarely weigh contextual fidelity. Instead, they amplify emotional tone, novelty, or virality. This distortion disincentivizes depth. Nuance, contradiction, and local specificity are often suppressed because they do not perform as well within recommender systems. The result is a homogenized feed: content that looks familiar, feels "on brand" and requires minimal interpretive effort. However, this streamlining comes at a cost. Without context, we lose our capacity to discern truth from parody, mistake exaggeration for fact, and struggle to evaluate intent.[5]

Nowhere is this more apparent than in political discourse. The need for algorithmic visibility now flattens posts that once relied on rhetorical strategy or insider references. When AI begins generating these posts, using pattern-matching rather than personal conviction, the distinction between authentic persuasion and synthetic mimicry blurs. Context collapse thus becomes not just a communicative risk but a civic hazard. Misinterpretation is no longer a byproduct of poor writing; it becomes embedded into the medium itself.[4]

For the human reader, this creates a form of cognitive exhaustion. Having to constantly parse whether a person or a machine wrote something, whether it is satire or sincerity, forces a state of perpetual ambiguity. In this fog, emotional responses take over. We react, scroll, and disengage. The work of meaning-making shifts to increasingly ill-equipped individuals to bear it, especially in information ecosystems that prioritize speed over verification.[6]

This collapse of context also undermines institutional credibility. When authoritative sources, such as news outlets, universities, and experts, appear alongside clickbait or AI-generated opinions, their words carry no distinct visual or structural difference. The trustworthiness of information is no longer derived from its source but from its format. Well-crafted language becomes the proxy for truth, regardless of who or what produced it. In such an environment, bad actors thrive. Bots that mimic real users, deepfake videos, and AI-generated misinformation gain traction precisely because the system no longer honors context as a protective layer.[3]

The Hybrid Mind must contend with this collapse not through nostalgia but through design. Restoring context does not mean resisting digital change, it means reinforcing human signals in systems that no longer default to it. This requires tools, governance, and cultural norms that elevate source transparency, authorial integrity, and audience awareness as active components of the content experience. The challenge is not just to consume content but to situate it. Moreover, in a synthetic world, situating meaning may be the most human task left.[10]

14.3   SYNTHETIC SYNDROME AT INTERNET SCALE

Synthetic Syndrome, initially introduced as a quiet psychological dislocation experienced by individuals working alongside generative AI, now expands into a larger societal condition. What once felt like an internal unease, imposter syndrome triggered by machine fluency, has become a systemic phenomenon reshaping entire industries. As generative AI continues to produce high volumes of convincing, coherent content, the roles traditionally reserved for human creativity, curation, and critical thinking begin to feel threatened not by replacement but by dilution.

In online journalism, for example, machine-generated summaries and AI-assisted headlines now populate major content platforms. The writer is no longer the primary voice but often an editor of algorithmically suggested material. This reconfiguration of authorship subtly demotes the act of writing from an original contribution to an evaluative task. Educators and students also report a similar shift. With tools that can generate essays, lesson plans, and even feedback, the pedagogical space becomes one of filtering rather than forming. Many report a strange hollowness, an erosion of meaning in an ecosystem where efficiency and adequacy replace effort and nuance.[4]

This condition, when scaled across professions, manifests as a cultural variant of Synthetic Syndrome. The absence of human input does not define it, but the presence of synthetic dominance. Creative professionals begin to question the value of producing work that can be mimicked in seconds. Designers encounter AI-generated templates that are "good enough" for clients. Coders face autocomplete engines that outpace them in speed but not in comprehension. The issue is not just job displacement; it is identity displacement. When the default content layer becomes synthetic, the human feels ornamental.[5]

Studies reinforce this observation. In a comparative analysis of bot versus human characteristics in online discourse, researchers found that bot-generated posts, particularly around trending events, tended to outpace human content in visibility and reach due to algorithmic alignment with platform norms.[3] Bots do not rest, hesitate, or second-guess. Their consistent presence creates a false consensus, a sense that certain viewpoints or behaviors are more prevalent than they are. This phenomenon compounds the emotional impact of Synthetic Syndrome by producing a world that feels increasingly disinterested in human irregularity or dissent.

The psychological cost becomes particularly acute in creative and meaning-centric roles. For many professionals, the sense of self is tightly coupled to output, what they write, build, or imagine. When machines offer rapid simulations of this output, often indistinguishable to outsiders, the foundational belief in human distinctiveness begins to waver. Some attempt to adapt by becoming AI supervisors or prompt engineers. Others withdraw. The deeper loss, however, is not economic but existential: a quiet erosion of meaning in doing the work once considered an expression of human essence.[11]

This is not merely a temporary phase of disruption. As the synthetic layer solidifies, new norms emerge around what constitutes legitimacy. If a grant proposal, student paper, or news article reads well, few interrogate its origin. The labor of meaning-making becomes invisible. Human quality becomes indistinct from machine-generated adequacy. The Hybrid Mind must recognize that Synthetic Syndrome, when scaled to the level of institutions, risks devaluing not just roles but the relational and affective labor that sustains communities of practice. What emerges is a future where presence is optional, and authenticity is a historical footnote.[10]

14.4   THE GOVERNANCE GAP

The proliferation of synthetic content across the internet has outpaced the development of tools to identify, regulate, or contextualize it. What results is a governance vacuum, an institutional blindness that leaves organizations, platforms, and citizens vulnerable to signal distortion, reputational risk, and decision-making based on unauthenticated input. This is not a failure of technology per se but a failure of coordination. While AI-generated outputs can be remarkably efficient, the mechanisms for accountability have lagged. Most platforms operate without a consistent standard for labeling machine-generated media, and few governments have passed enforceable legislation requiring provenance tracing at scale.[1]

Watermarking has been proposed as a solution to this crisis of authenticity. In theory, digital watermarks embedded in AI-generated images or text can help distinguish synthetic from human-authored content. OpenAI, for example, has begun integrating C2PA-compliant watermarks into DALL·E 3 outputs, marking a step toward traceable authorship in generative imagery. The Content Authenticity Initiative (CAI), a coalition including Adobe and major media entities, has made similar efforts to embed secure provenance metadata into digital content.[5] However, adoption remains voluntary, fragmented, and technically fragile. Watermarks can be removed, bypassed, or rendered unreadable when content is modified, even inadvertently.

Legal frameworks have begun to stir. California's AB 3211, a first-of-its-kind watermarking bill, mandates that AI-generated audiovisual content carry metadata indicating its origin. Nevertheless, enforcement is minimal, and platforms can still circulate altered or repackaged media that escapes detection. The same systems that enabled generative content at scale, deep learning, content distribution networks, and recommender algorithms were not built with traceability in mind. Retrofitting them now involves systemic overhaul and cross-institutional cooperation that has not yet materialized.[6]

This lack of traceability leads to profound governance challenges. Without clear labels or trustworthy authentication protocols, humans are forced to navigate a sea of ambiguous information. The cost is cognitive, reputational, and civic. Misinformation campaigns become easier to execute and harder to trace. Legal disputes around ownership or defamation become murkier. Democratic institutions face manipulation from synthetic actors that cannot be easily identified, let alone held accountable for their actions. In such an environment, even well-intentioned actors may inadvertently amplify false signals.[7]

From a strategic innovation standpoint, the absence of evaluative metrics for synthetic content integrity is a glaring omission. The ECTM, previously introduced as a framework for innovation governance, can be extended to this domain by scoring content ecosystems across dimensions such as authenticity risk, traceability readiness, and provenance assurance. These metrics would not merely assess technology but would help organizations prioritize ethical content pipelines and invest in infrastructure that supports validation.[1] Just as cybersecurity practices become essential to digital infrastructure, authentication protocols must become a pillar of information systems.

The Hybrid Mind cannot assume that platforms will self-correct. In a market optimized for scale and engagement, synthetic content thrives precisely because it performs. The burden falls to institutional actors, governments, educators, technologists, to design for discernment.[8] Whether through watermarking mandates, educational campaigns, or algorithmic transparency requirements, governance must evolve to meet the reality of a synthetic world. Otherwise, we risk not only misinformation but also the systematic erosion of signal fidelity, the very foundation of collective sense-making.[9]

14.5   RE-AUTHENTICATION AND HUMAN PRESENCE

In a world increasingly shaped by synthetic information, reestablishing human presence becomes a design imperative, not just a philosophical stance. The digital commons no longer offer any inherent guarantee that what we see, hear, or read originates from another human being. This erasure of presence undermines our ability to relate, trust, and discern meaning. The task before us is not to reject technological augmentation but to build architectures that center human intentionality, making it visible, verifiable, and valuable again.

Re-authentication is the cornerstone of this effort. It refers to the intentional reinsertion of provenance, context, and authorship into digital content. Initiatives such as the Content Authenticity Initiative (CAI) and the Coalition for Content Provenance and Authenticity (C2PA) have laid the technical groundwork for this shift. By embedding metadata, digital signatures that track the origin and modification history of media, these frameworks provide a blueprint for regaining trust in the face of synthetic saturation.[5] OpenAI's integration of C2PA watermarking in DALL·E 3 demonstrates that even large-scale generative platforms can take steps toward attribution. However, challenges remain in terms of watermark resilience and consistency of adoption.[6]

Nevertheless, authentication alone is not enough. The Hybrid Mind must think beyond traceability toward design principles that restore friction, deliberate slowness, as a signal of human thought. Instant publishing, zero-click interaction, and stream-based consumption reinforce shallow engagement. Meaning, by contrast, often requires effort, ambiguity, and reflection. Platforms can counter synthetic acceleration with interface rituals that slow users down in meaningful moments. For instance, "meaning checkpoints" could prompt creators to add context or require readers to confirm their interpretation before resharing controversial content. These gestures are small but mighty: they assert that interpretation is not a commodity but a shared responsibility.[11]

This philosophy also extends to personal practice. Digital detoxes, once seen as trendy wellness rituals, are becoming critical strategies for restoring cognitive sovereignty. Empirical studies have demonstrated that even brief breaks from social media result in measurable improvements in mental clarity, sleep quality, and emotional regulation.[12] In an age of ambient automation, stepping away from the algorithmic stream is not escapism, it is a form of resistance. It provides the psychic space to recognize one's thoughts amid the hum of machine-generated noise. As such, detoxification should not be framed as abstention but as recalibration, a way to restore rhythm, focus, and presence.

Designing the future net also means foregrounding presence in collaborative systems. Co-creation platforms that enable users to distinguish between AI contributions and human edits, for example, empower users to take agency over the creative process. Hybrid interfaces, those that display provenance layers, source hierarchies, or inference transparency, help users understand how knowledge is constructed, not just what it concludes. These tools are not merely informational. They affirm the dignity of participation. They remind us that to create is to take a position, to leave a lasting mark, and to consider the consequences.[10]

Ultimately, re-authentication is not about defending the past. It is about safeguarding the conditions that make meaning possible in the present. In the noise of simulation, presence becomes a rare currency. The Hybrid Mind does not retreat from this challenge. It moves toward it, curating architectures, rituals, and tools that protect the human core. In doing so, it builds a world where being real is not just verifiable but vital.

CHAPTER SUMMARY

The rise of the synthetic internet has revealed more than a technological inflection point; it has exposed a deeper cultural and epistemic shift. As synthetic content becomes the default substrate of digital life, humans must grapple with the collapse of context,[2] the scaling of Synthetic Syndrome,[5] and the widening governance gap.[1] Meaning no longer arrives through proximity or authorship, it must be reclaimed through design, ritual, and deliberate presence.[11] Re-authentication emerges as a survival instinct for the Hybrid Mind, not merely to verify what is real but to preserve the conditions under which reality still matters.[10]

Nevertheless, the challenge is not simply defensive. It is also creative. The next frontier for the Hybrid Mind is not just recognizing the synthetic world, but shaping it. In the following chapter, we move from crisis to cultivation. Hybrid Minds in Culture explores how human-AI partnerships are already generating new forms of expression, storytelling, and beauty. Where this chapter diagnoses signal loss, the next reveals cultural emergence: art born of the algorithm, music rooted in augmentation, and heritage reimagined through code. If Chapter 14 asks how we preserve what is human, Chapter 15 asks how we evolve it.

REFERENCES

1. Joseph X. Ng, *Data Insights: Core Principles of Statistical Analysis for Storytelling* (SAGE Ensights LLC, 2024), https://a.co/d/edWAFvL.
   Ng provides a foundation for understanding how data loses its narrative coherence in synthetic environments, reinforcing the chapter's focus on the erosion of interpretive anchors in high-volume, AI-generated content.
2. Cathy O'Neil, *Weapons of Math Destruction: How Big Data Increases Inequality and Threatens Democracy* (Crown Publishing, 2016).
   O'Neil's examination of opaque algorithmic systems supports the chapter's argument that synthetic content ecosystems replicate patterns of invisibility, harm, and unaccountable influence.
3. Joseph X. Ng and Andres Fortino, "Strategic Innovation Governance: Identifying Crucial Use Cases via Experimentum Crucis Technology Matrix (ECTM)" (2025).
   This governance model underpins the chapter's call for rigorous frameworks to score the ethical and traceability risk of synthetic content systems, linking policy to platform-level implementation.
4. Gartner, *Why AI Pilot Projects Fail* (Gartner Research, 2024).
   The report provides organizational context for the widespread lack of AI governance, explaining how unchecked content generation scales without accountability or metrics for meaning and provenance.
5. Sherry Turkle, *Reclaiming Conversation: The Power of Talk in a Digital Age* (Penguin Press, 2015).
   Turkle's research illuminates how digital interaction design contributes to emotional disconnection and flattening of communication, validating the chapter's critique of synthetic saturation as a barrier to presence and meaning.
6. Sandra Wachter and Brent Mittelstadt, "A Right to Reasonable Inferences: Re-thinking Data Protection Law in the Age of Big Data and AI," *Columbia Business Law Review* 2 (2019): 494–620.
   This article strengthens the chapter's advocacy for re-authentication, offering legal and ethical support for transparency in AI-generated content systems.
7. European Commission, *Ethics Guidelines for Trustworthy AI* (2019).
   These guidelines anchor the chapter's emphasis on accountability and traceability as prerequisites for ethical synthetic media governance.
8. European Commission, *Artificial Intelligence Act Proposal* (2021).
   The proposal's tiered approach to risk aligns with the chapter's framework for categorizing and regulating synthetic content based on its potential to distort public discourse or identity.
9. National Institute of Standards and Technology (NIST), *AI Risk Management Framework (AI RMF 1.0)* (2023).
   The NIST model is used in the chapter to suggest how provenance, watermarking, and auditability should be integrated into platform-level risk mitigation efforts.
10. Floridi, Luciano, and Josh Cowls. "A Unified Framework of Five Principles for AI in Society." *Nature Machine Intelligence* 1, no. 1 (2019): 65–67.
    This framework provides normative grounding for the chapter's emphasis on human agency and explicability in content ecosystems overrun by synthetic media.
11. Fei-Fei Li, "Human-Centered AI: Building Trust with Empathy and Ethics," *Stanford HAI Lecture* (2021).

Li's lecture supports the chapter's thesis that systems must be designed to preserve human intention, identity, and context – especially as synthetic outputs become the dominant form of digital expression.

12. Brown, Lauren, and Daria Kuss. "The Effects of Partaking in a Two-Week Social Media Digital Detox: A Mixed Methods Study." *International Journal of Environmental Research and Public Health* 20, no. 2 (2023): 1206. https://doi.org/10.3390/ijerph2 0021206.

This study offers empirical backing for the chapter's proposal that deliberate breaks and interface frictions can restore cognitive clarity and reduce synthetic-induced fatigue.

15 Hybrid Minds in Culture

The Human-AI Convergence delves into the dynamic relationship between culture and intelligence systems, with a focus on the pivotal role of the Hybrid Mind. This chapter redefines the Hybrid Mind, not just as a strategic or cognitive archetype but as a cultural agent. Throughout history, culture has served as a mirror and a mechanism, reflecting the spirit of its time while guiding collective memory, values, and expression. The advent of generative AI has deepened this relationship. Machines are no longer passive conduits of culture; they now generate, curate, and sometimes reinterpret cultural meaning. This chapter explores how these shifts are shaping new forms of co-creation, preservation, and emotional experience in art, literature, music, and heritage.

We begin by examining the emergence of a hybrid aesthetic, a creative language formed at the intersection of human intent and machine-generated content. AI-generated works, ranging from images to orchestral compositions, are being showcased in galleries, concert halls, and publishing platforms. Nevertheless, this wave of creativity comes with conceptual tension. The Hybrid Mind must navigate a dual awareness: that while machines can replicate structure and style, they do not possess memory, intent, or lived experience. Hybrid cultural artifacts are thus not valued solely for their novelty but for the ways they maintain human resonance. Artists like Sougwen Chung, who blend human movement with robotic gestures in real-time performance, demonstrate how collaborative practice can extend rather than replace the emotional dimension of creative work.[1]

The chapter then shifts to the crucial role of the Hybrid Mind in cultural preservation, a domain where AI contributes significantly to the safeguarding of collective memory. From 3D scanning sacred sites through projects like CyArk to reconstructing endangered languages with neural networks, algorithms are now being leveraged to prevent the erosion of cultural knowledge. These efforts, however, must be guided by ethical and contextual understanding. Without community participation, preservation becomes appropriation. The Hybrid Mind, acting as a cultural steward, insists on fidelity, not just to data, but to people. This requires rethinking AI development as a cross-cultural dialogue, not a technical export.[2]

The psychological consequences of automation are also explored. Synthetic Syndrome, a term developed to describe the erosion of creative confidence in the

DOI: 10.1201/9781003648338-20

presence of high-performing AI, reveals how identity, pride, and authorship are impacted when machines enter spaces once considered uniquely human.[3] This syndrome does not stem from failure but from success, where human creators fulfill their task, only to question its worth once they realize it could have been automated. The Hybrid Mind must absorb this emotional signal and respond by designing systems that center the human, not in function, but in feeling.

Finally, the chapter reframes culture itself as infrastructure. Culture is not just art and artifacts, it is the unseen protocol by which values are encoded into systems. As AI technologies are deployed across different societies, their design must be sensitive to the symbolic and moral frameworks of those communities. UNESCO's ethics recommendation for AI development emphasizes the importance of respecting cultural pluralism, thereby challenging the presumption of universality in machine learning logic.[4] The hybrid Mind, acting as translator and mediator, plays a pivotal role in ensuring that AI not only scales globally but remains locally meaningful.

In sum, this chapter positions Hybrid Minds as cultural architects, individuals who collaborate with machines not only to generate expression but to preserve identity, evoke emotion, and encode values. In doing so, they transform AI from an engine of automation into an instrument of cultural continuity and renewal, ensuring that our cultural identity remains intact and connected across generations.

15.1 THE CULTURAL EMERGENCE OF THE HYBRID AESTHETIC

Culture is not merely the output of a society's imagination, it is its nervous system, registering shifts in values, meaning, and identity. As AI systems increasingly participate in the creation of cultural artifacts, such as art, music, literature, and even design, we begin to witness the emergence of what could be termed a hybrid aesthetic. This is not a passing novelty but a realignment of creative authorship. In this new paradigm, generative models such as DALL-E, Midjourney, and ChatGPT act not as mere tools but as semi-autonomous collaborators. Their contributions, though algorithmic, increasingly bear the illusion of intent. When a generative model produces a landscape that evokes serenity or writes a sonnet that triggers nostalgia, the line between imitation and expression starts to blur. This unsettles traditional definitions of artistry.

However, the presence of style is not the same as the presence of soul. Aesthetic coherence does not imply creative intuition. Furthermore, it is at this point that the Hybrid Mind must intervene. The role of the human in this equation is not to outdraw the AI but to encode a deeper purpose, to use machine outputs as mirrors or launchpads for emotional resonance. The real question is not "Can AI create art?" but "Can humans still recognize themselves in AI-assisted creation?" That distinction is crucial for cultural continuity.

This emergence is not isolated. It finds resonance in the work of artists like Sougwen Chung, whose performances merge robotic gestures and human movement to create visual improvisation that transcends authorship.[1] Her work exemplifies the hybrid aesthetic not as a compromise but as an expansion, where machines do not substitute for creativity but amplify its scale and reach. This shift marks a more profound transformation: the Hybrid Mind as an artist-curator, one who innovates and empowers the evolution of the hybrid aesthetic.

Nevertheless, what makes this moment distinct is not simply the presence of machine creativity, it is the acceleration of cultural change. Hybrid aesthetics evolve in tandem with software updates. New versions of generative models introduce novel capabilities, styles, and techniques at rates that outpace traditional artistic schooling. A once-niche technique today may be widely adopted by the next release. This rapid diffusion flattens stylistic hierarchy and challenges traditional forms of mastery. For emerging creators, the hybrid aesthetic democratizes access to tools and resources, making them more accessible. For established practitioners, it demands a redefinition of identity, urging them to adapt and innovate in the face of rapid change.

The Hybrid Mind, in this context, is not a philosophical metaphor, it is a necessary posture. It is the cognitive and emotional stance we adopt when navigating a world in which machines increasingly participate in creating meaning. It is the refusal to collapse into either technophobia or techno-utopia. Instead, it is the willingness to treat cultural systems as collaborative environments where human intent and machine capacity co-evolve. This is not about conceding creativity to machines. It is about reclaiming the human signature in an environment where authorship is ambient and distributed. We do not ask whether the machine can create, we ask whether the creation still reflects our questions, our hopes, our contradictions.

If culture is how we tell the story of being human, then the hybrid aesthetic is a new chapter in that story. However, it is one we must write consciously, with both our hands and our machines on the pen.

15.2   HERITAGE THROUGH ALGORITHMS

Culture is not just created, it is remembered. Moreover, memory, at scale, is becoming an increasingly technical challenge. Across the world, communities are racing against time and climate and neglect to preserve their traditions, languages, and artifacts. In this urgent context, artificial intelligence has emerged not only as a means of generating new cultural forms but also as a partner in safeguarding the old. The Hybrid Mind extends its role here, not as a passive receiver of history, but as an active restorer. AI-driven techniques in digitization, transcription, and pattern reconstruction are being deployed to capture and simulate cultural expressions that might otherwise be lost to erosion, displacement, or disaster.

Projects like CyArk exemplify this shift in preservation strategy. CyArk utilizes AI-enhanced 3D scanning and photogrammetry to digitally archive sacred and endangered sites worldwide, from the temples of Angkor Wat to the churches of Lalibela. These digital twins are not simply records; they are interactive blueprints that can support education, tourism, and even reconstruction after a catastrophe.[2] AI plays a crucial role in translating these scans into coherent models, detecting damage, and even predicting structural vulnerabilities. However, it is the human historian, anthropologist, and architect who assigns meaning to the data. This is the Hybrid Mind in action, where machine precision complements cultural intention.

Beyond physical heritage, algorithms are also playing a role in the recovery of linguistic and musical heritage. AI systems trained on fragmented audio or text are now helping linguists revive lost dialects, reconstruct oral histories, and simulate traditional musical patterns that have no surviving performers. These acts of cultural recovery extend far beyond mere novelty; they are acts of justice. For many marginalized or colonized communities, cultural loss is not accidental, it is the residue of systemic erasure. In these cases, algorithmic restoration must be designed with humility and collaboration. The Hybrid Mind does not just ask what can be preserved, but who is empowered in the process.

Researchers have proposed the use of Unique Authenticity Identifiers (UAIs) for artworks and crafts, not only to prevent forgery but to ensure recognition of original creators, particularly in Indigenous and diasporic communities.

Still, this partnership is not without risk. When algorithms are trained on cultural datasets without transparency, the act of preservation can quickly slide into appropriation. Without the consent or involvement of origin communities, well-meaning projects can replicate the same extractive patterns they claim to remedy. This is why Hybrid Minds are needed, not just coders or curators but facilitators of ethical collaboration between communities and machines. AI cannot hold memory in the way humans do, it lacks reverence, lineage, and grief. However, it can assist those who do. The Hybrid Mind ensures that memory is not just stored but honored.

The shift toward algorithmic preservation is not about outsourcing history to machines. It is about extending the reach of those working to protect it, ensuring that tomorrow's generations inherit more than artifacts; they inherit access, understanding, and a sense of pride.

15.3 SYNTHETIC SYNDROME AND THE EMOTIONAL TOLL OF CULTURAL AUTOMATION

As artificial intelligence continues to reshape creative industries, a more subtle consequence has surfaced, what we call Synthetic Syndrome. This phenomenon describes the psychological and emotional friction experienced by artists, writers, musicians, and other creatives when their sense of identity and value is disrupted by the increasing fluency of AI in cultural domains. Unlike traditional automation, which often replaces repetitive or physical labor, AI's intrusion into expressive work directly challenges how individuals define purpose, meaning, and mastery. For those who have spent years honing their craft, the realization that a model can generate something indistinguishable in seconds can induce not only economic anxiety but a profound existential disorientation.

This is not irrational fear; it is a legitimate response to a shift in authorship. While AI-generated and human-created artworks can appear similar in technical quality, people consistently find human works to be more emotionally resonant and meaningful, suggesting that authenticity and intention still shape how we connect with creative expression. This gap reveals something critical: humans evaluate art not solely by aesthetic criteria but by the perceived authenticity of its origin. When the origin becomes ambiguous, so too does the emotional weight of the work. For creators, this erosion of meaning can lead to a decline in motivation, imposter syndrome, and even withdrawal from their practice.

The Hybrid Mind does not simply accept this as collateral damage. It treats it as an ethical imperative. The goal is not to restore a myth of creative exclusivity but to reestablish emotional equilibrium. This requires the development of tools, frameworks, and feedback systems that preserve the human role, not merely as a curator of outputs but as a meaning-making agent. Emotional ergonomics becomes a priority. Just as physical tools were designed to reduce repetitive strain, AI systems must be designed to reduce emotional displacement. Features such as attribution traceability, human-in-the-loop co-creation, and contextual prompts can help restore a sense of participation and authorship.

The cultural industries must, therefore, move beyond efficiency metrics and begin to account for the affective labor of creation. We must ask not just whether AI-generated stories or paintings are "good" but whether they contribute to or detract from the inner lives of human creators. A Hybrid Mind culture respects both the output and the process, building systems that sustain emotional authenticity rather than replacing it with procedural fluency. Meaning is not a byproduct of creation; it is its foundation. Moreover, when the foundation cracks, even the most compelling artifacts become weightless.

Synthetic Syndrome is not a pathology to be treated. It is a signal to be understood. It reminds us that culture is not only what we produce but how we feel while producing it.[3]

15.4   THE ROLE OF AWE IN HYBRID CREATIVITY

In the conversation about artificial intelligence and culture, efficiency, productivity, and fidelity often dominate the discourse. However, beneath these metrics lies a less quantifiable yet profoundly human dimension, awe. Awe is not a technical attribute. It is not something we optimize or benchmark. It is the sudden expansion of perception, the moment of stillness when confronted with beauty, complexity, or the sublime. In cultural creation, awe is what separates a technically impressive piece from a transformative experience. As AI systems become partners in the creative process, the role of the Hybrid Mind is to preserve and amplify this sense of wonder, not to automate it away.

Awe operates at the intersection of emotion and cognition. It slows time, elevates meaning, and connects the viewer or listener to something larger than themselves. While machines can now compose orchestral scores or produce paintings with uncanny realism, awe is not a guaranteed output of skill or fidelity. It emerges from intention, ambiguity, surprise, and resonance. Hybrid creativity, where human sensibility guides AI output, can evoke awe when the collaboration transcends expectation. A human mind, steering a generative model not toward replication but revelation, crafts an experience where the unfamiliar still feels intimate.

The artist Sougwen Chung provides a powerful example. In her collaborative performances with robotic drawing arms, the choreography between human movement and machine response becomes a meditation on coexistence. The audience is not just watching art, they are witnessing a conversation across boundaries. Chung does not merely use AI for efficiency. She invites it into ritual, creating space for reflection, silence, and even spiritual tension.[1] Her work exemplifies the Hybrid Mind at its most evocative, where the machine is not the spectacle but the spark.

This framing challenges dominant narratives of AI in art. Much of the public discourse remains fixated on realism, whether an AI-generated image appears to be a photo or whether AI-generated music sounds indistinguishable from that of a human composer. However, realism is not the same as resonance. A painting that mimics the texture of oil or a symphony that follows harmonic rules may impress, but it does not necessarily move the listener. The Hybrid Mind recalibrates this metric. The measure is not fidelity to form but fidelity to feeling.

In many traditions, religious, philosophical, and artistic, awe is a necessary threshold for wisdom. It invites humility, reflection, and a sense of participation in something enduring. If AI is to contribute meaningfully to culture, it must be guided by human agents who understand this threshold. The Hybrid Mind, then, is not only a systems thinker but a steward of awe. It ensures that even in a world of automation, we do not lose our capacity to be stopped by beauty or stirred by mystery.

As algorithms continue to scale, we may become one of the few remaining signals of authenticity. Not because machines cannot trigger it, but because only humans can know when it matters.

15.5   FROM CULTURAL OBJECTS TO CULTURAL OPERATING SYSTEMS

Culture is more than expression, it is infrastructure. While we often encounter culture as an object or performance, its deeper function is systemic: it determines what is valued, how meaning is negotiated, and how norms are sustained. In the age of AI, culture now shapes how machines are trained, how fairness is encoded, and how adoption unfolds across geographies. The Hybrid Mind, in its cultural role, is not just a creator of artifacts, it is a systems architect responsible for embedding cultural awareness into the very logic of intelligent systems. Culture is no longer only downstream of technology; it is upstream of design.

This reframing invites us to think of culture not as a product but as a protocol. Just as operating systems define the rules by which software runs, cultural operating systems influence how AI behaves in different contexts. Consider the deployment of language models in multilingual societies. A system trained predominantly on English-language data will carry not only linguistic biases but also cultural ones, norms around politeness, directness, humor, and hierarchy. These invisible values can easily misfire when exported without adaptation. The Hybrid Mind understands that translation is never just about words, it is about worldviews.

Initiatives such as UNESCO's AI ethics framework have emphasized this point, advocating for pluralism in the development and deployment of AI. Their position is clear: global AI systems must respect local epistemologies, traditions, and values, or risk becoming instruments of cultural domination.[4] In practice, this means including community leaders, anthropologists, and ethicists in model development. It means training AI not just on diverse data but with contextual constraints. The Hybrid Mind becomes a translator, of values, of intentions, of consequences. It is a buffer against flattening forces.

A telling example is found in content moderation. What constitutes hate speech in one culture may be considered satire in another. Automated filters trained on Western norms often fail in indigenous, diasporic, or non-Western settings. This is not just a technical problem, it is a cultural one. Without cultural operating systems to mediate meaning, machine systems will continue to misclassify, silence, or amplify in ways that distort local realities. The Hybrid Mind must anticipate this and design for contextual nuance rather than making a universal assumption.

This perspective also opens the door to intentional hybridity in AI training, where models incorporate symbolic logic from diverse world philosophies, ethical schemas rooted in collectivist or relational paradigms, and decision-making structures that reflect spiritual or ecological priorities. These are not academic fantasies, they are design challenges. Furthermore, they will determine whether AI is merely globally accessible or truly globally relevant.

In this emerging era, the Hybrid Mind is not just a cultural consumer, it is a cultural engineer. It does not replicate what already exists. It codes what could be. In doing so, it preserves the richness of human plurality within systems that might otherwise homogenize. When we treat culture as an operating system, we stop asking only what AI can do; instead, we ask what AI can do for us. We start asking: Under whose logic, in whose voice, and toward what future?

CHAPTER SUMMARY

The emergence of Hybrid Minds within cultural domains signals a more profound transformation than mere technological disruption; it reflects a new way of creating, preserving, and interpreting meaning in an increasingly automated world. As we have seen, AI is no longer relegated to technical utility; it now co-authors the stories, symbols, and memories that define collective identity. Whether through collaborative art,[1] algorithmic preservation of endangered heritage,[2] or the emotional reckoning of Synthetic Syndrome,[3] human-AI interaction in culture forces us to ask: What does it mean to create, and who gets to claim authorship?

This chapter affirmed that awe remains a central metric for human authenticity in the age of generative systems.[1] The Hybrid Mind does not seek to outpace AI but to guide it, to choreograph creativity, to design for emotional resonance, and to embed cultural fluency into the protocols of machine intelligence. By treating culture as a kind of operating system, the Hybrid Mind assumes responsibility not only for the outputs of AI but also for the worldviews and values that those outputs encode.

As we turn to the final chapter, we move from reflection to proposition. If Hybrid Minds are creators, curators, and stewards, then they are also rights holders. The next chapter asks: What ethical, cognitive, and emotional rights must be protected as humans increasingly interact with intelligent systems? In the same way that societies once defined civil liberties for industrial labor, we now face the challenge of articulating the rights of the Hybrid Mind. This is not just a legal concern, it is a cultural one. Because to protect what is human, we must first agree on what is worth preserving.

REFERENCES

1. *TIME Magazine.* "Sougwen Chung's Robotic Improvisation." *TIME*, 2024.
    This article highlights Sougwen Chung's work at the intersection of robotics and performance art. Her process exemplifies the emergence of hybrid aesthetics, where human intent and machine behavior converge to create emotionally resonant cultural expressions – a central theme in this chapter.
2. Wikipedia Contributors. "CyArk and Digital Heritage Preservation." *Wikipedia*, 2024.
    CyArk's efforts to digitize endangered heritage sites serve as a real-world example of AI-enabled cultural preservation. The project underscores how machines can aid – but not replace – human stewardship of collective memory.
3. Ng, Joseph X. "Synthetic Syndrome: The Silent Impact of AI on Human Confidence" [White paper], LMX Partners, 2025.
    This white paper formally introduces the concept of Synthetic Syndrome. It anchors the chapter's psychological exploration of identity loss and creative detachment in AI-augmented environments.
4. Ng, Joseph X. *Data Insights: Core Principles of Statistical Analysis for Storytelling* (New York: SAGE Ensights LLC, 2024).
    Ng's book illustrates how narrative interpretation of data enhances human meaning-making, reinforcing the chapter's argument that storytelling remains a fundamentally human act – even in AI-augmented cultural systems.

Conclusion

Ultimately, the Hybrid Mind is not a theory or a trend. It is a lived condition. To be human in the age of artificial intelligence is to hold the tension between acceleration and meaning, between automation and presence, and between the efficiency of machines and the vulnerability of human judgment. The work of this book has been to name, frame, and design for that condition, to provide language for what people are already experiencing and to offer structure for what must now be led with intention.

We began by redefining intelligence itself. No longer confined to either human brilliance or machine calculation, intelligence is now hybrid, emerging from the interplay of intuition, context, pattern recognition, and feedback. From this interplay, new responsibilities arise. Humans can no longer afford to see themselves only as operators of tools. The role has changed. We are now orchestrators of systems, curators of meaning, and stewards of ethical hesitation in environments built for speed. This evolution from task execution to sensemaking is not optional. It is the cost of coherence in a world run by agents who do not sleep, second-guess, or suffer. Only we do that. Moreover, that is why we matter.[1]

The architecture of intelligence is more than code. It shapes how we feel, decide, and connect. As we develop agentic systems and modular models, we must recognize that design decisions are inherently moral decisions. When feedback is absent, systems drift. When modularity is ignored, complexity collapses. The architecture must reflect our values, not just our ambitions. It must remember what is hard to measure, presence, trust, surprise, and the right to slow down. Without these, intelligence becomes performative, and governance becomes retrospective instead of generative.[2]

Across sectors, the question is no longer whether AI will change work, society, or our sense of self. It already has. The question is whether we will lead that change or become its artifacts. Governance must not wait for collapse to trigger reform. The Experimentum Crucis Technology Matrix (ECTM) and the Hybrid Audit Model (HAM) were introduced as living structures for this leadership, tools to ensure we are designing for alignment, not just accuracy. They remind us that the most advanced systems are only as sound as the questions we are still brave enough to ask. Without reflection built into the loop, innovation becomes inertia. Forward motion, without direction, is not progress, it is drift.[3]

To lead well, we must also collaborate well, not just with people but with systems that now have voice, initiative, and presence. The politics of that presence, who gets to decide, to translate, to be seen, are shaping the future of co-creation. Human-centered collaboration demands more than user-friendly interfaces. It demands cognitive and cultural humility. It requires us to notice the invisible labor of interpretation, the emotional cost of proximity to machine fluency, and the long arc of trust that cannot be automated. This is the quiet work of Hybrid Minds, those who carry both the spark and the skepticism, the systems and the soul.[4]

And then there is the future, still unwritten but already in play. The Hybrid Mind does not see the future as something to survive. It sees it as something to shape. That shaping will require new rights, new rituals, and new rhythms of resistance against speed for speed's sake. It will require that we remember what it feels like to think slowly, make deliberate decisions, and design with consequences. Forward Humanity is not about being first. It is about being grounded. Furthermore, that, perhaps, is the most important convergence of all, not between human and machine, but between capacity and care, between capability and conscience.[5]

REFERENCES

1. Joseph X. Ng, *Data Insights: Core Principles of Statistical Analysis for Storytelling* (SAGE Ensights LLC, 2024), https://a.co/d/edWAFvL.
    Ng's work emphasizes the use of statistical storytelling to bridge data and decision-making. It supports the book's final reflection on how hybrid intelligence must retain interpretive humility and narrative clarity in algorithmic environments.
2. Fei-Fei Li, "Human-Centered AI: Building Trust with Empathy and Ethics," *Stanford HAI Lecture* (2021).
    Li's lecture emphasizes that trust in AI must be built through empathetic design. Her perspective informs the book's Conclusion by reaffirming that augmentation must enhance human dignity, not merely automate cognition.
3. Joseph X. Ng and Andres Fortino, "Strategic Innovation Governance: Identifying Crucial Use Cases via Experimentum Crucis Technology Matrix (ECTM)" (2025).
    This paper introduces the ECTM framework, which serves as the conceptual foundation for much of the book's strategic innovation governance model. In the Conclusion, its logic underpins the call for rigorous, value-driven technology scaling aligned to enterprise-level impact.
4. Joseph X. Ng, *Mastering IT Innovation: 90-Day Proof of Value Blueprint* (SAGE Ensights LLC, 2024), https://a.co/d/07ZO4zW.
    Ng's blueprint supports the Conclusion's emphasis on iterative validation and responsible scaling. It reinforces that innovation maturity requires both governance and agility in tandem.
5. Yuval N. Harari. *Homo Deus: A Brief History of Tomorrow* (Harper, 2017).
    Harari's speculation on humanity's future amid intelligent systems is used to reflect on the existential and philosophical stakes of the Hybrid Mind era, questioning what remains essentially human as our tools become intelligent.

Epilogue

BEYOND THE LOOP: THE MOMENT IT BECOMES A THANG

We rarely notice the threshold until we are already standing on the other side. That is the curious thing about transformation. It does not always come with announcements or user manuals. More often, it arrives disguised as convenience, an AI suggestion, an automated reply, a dashboard insight we did not ask for but somehow trust. It happens quietly until suddenly, it is everywhere. Moreover, that is the essence of the phrase: "Not a thang until it becomes a thang." It is not just a clever refrain; it is an astute observation about tipping points. What was once fringe becomes foundational, and what felt optional becomes obvious.[1]

The Hybrid Mind is not a forecast. It reflects our now. Systems of augmentation are no longer locked in research labs. They have infiltrated our meetings, classrooms, creative tools, and operating procedures. The quiet revolution is already underway. AI does not just write, summarize, or predict, it shapes tempo, suggests meaning, and reframes relevance. But with that, something subtle shifts in us, too. Many describe a kind of dissonance, a feeling of being near brilliance yet slightly displaced by it. This phenomenon, which we have called Synthetic Syndrome, is the emotional residue of an era in which the machine appears to anticipate our next move. Synthetic Syndrome is the feeling of being both empowered and overshadowed by the capabilities of AI, a feeling that is becoming increasingly common in our AI-driven world.[1]

In this convergence, the tipping point is not technological, it is human. The moment something becomes a thang, it does so not just because it works but because people integrate it into meaning-making. They build trust, habits, and identities around it. A model that interprets radiology scans or drafts legislation does not just execute; it becomes part of the human decision environment. The Hybrid Mind, as a framework, exists to make this visible. It argues that in every synthetic suggestion lies an opportunity for human reinterpretation. The real differentiator is not speed or scale, it is our ability to decide what still matters and why.[1]

Leadership in this new era demands more than fluency in platforms. It requires clarity of presence. The Experimentum Crucis Technology Matrix (ECTM) was introduced to guide this discernment. It is not enough to build what is possible; we must decide what is permissible, what is ethical, and what is worthy of our attention and trust. ECTM offers a structured approach to evaluating use cases, not only in terms of feasibility but also in terms of strategic alignment, human dignity, and long-term cultural impact.[2] Alongside this, the EXAI Manifesto advocates for lifecycle responsibility, where accountability is maintained from the moment a system is conceived to the moment it is retired.[1] Together, these governance models do not just set standards, they protect space for human judgment.

Furthermore, feedback, often treated as a technical function, is reframed here as the heartbeat of ethical collaboration. The OODA-F loop, with its formal recognition of feedback as a distinct and dynamic phase, reminds us of that action without reflection is acceleration without alignment.[2] Feedback in hybrid systems is not just about whether something worked, it is about how it felt, whom it affected, and what unintended dynamics it produced. In a decision loop shared by humans and machines, feedback must become not just telemetry but interpretation. When humans lose their place in the loop, they do not disappear; they become invisible. They drift. The job of design, then, is to re-center them.

One final story, then. Imagine a teacher preparing lesson plans. She uses an AI tool that suggests modules, assessments, and even student engagement strategies. It is efficient and helpful. However, she notices something. Her lessons feel overly polished. Her student feedback starts sounding alike. The creative messiness she once embraced – the improvisation, the cultural references, the spontaneous dialogue – is more complex to locate. So, she pauses. She rewrites a unit by hand. She brings in a poem that is not part of the system's recommendations. Her students respond. Not because the AI failed but because she chose to be present. In that moment, the Hybrid Mind becomes more than a framework. It becomes a posture of presence, a decision to lead, not just with intelligence, but with meaning.

REFERENCES

1. Joseph X. Ng, "EXAI Manifesto: Ethical eXcellence in Artificial Intelligence," Humainority.org (2025), https://Humainority.org/exai-manifesto/.
    The EXAI Manifesto distills the Epilogue's overarching appeal into a practical charter for ethical, accountable, and human-centered AI development. It functions as a moral blueprint for the Hybrid Mind paradigm.
2. Joseph X. Ng and Andres Fortino, "Strategic Innovation Governance: Identifying Crucial Use Cases via Experimentum Crucis Technology Matrix (ECTM)" (2025).
    This paper introduces the ECTM framework, which underpins the book's call in the Epilogue for discerning between novelty and necessity when scaling AI systems. It offers a governance lens for responsibly navigating technological convergence.

Acknowledgments

First and foremost, I offer my deepest gratitude to my parents, whose strength, sacrifice, and enduring values have shaped who I am. To Matthew, thank you for your constant encouragement, steady wisdom, and grounding presence throughout this journey. To my sons, your curiosity, joy, and boundless imagination are daily reminders of why the future we build matters. This work is, in many ways, for you. And to Tasha, thank you for your unwavering patience, support, and belief in this vision. Your presence has been my anchor and my inspiration.

To the many colleagues and collaborators across my career, your shared vision, insightful challenges, and enduring partnership have not only profoundly influenced the perspectives and principles in this work but also made it richer and more comprehensive. I am especially grateful to my colleague at NYU, Dr. Andres Fortino, for his unwavering advice, mentorship, and co-authorship of the Experimentum Crucis Technology Matrix (ECTM). His strategic insight and academic rigor have helped shape this framework into a cornerstone of responsible innovation governance.

To my students at NYU, thank you for your curiosity, engagement, and relentless pursuit of insight. Your contributions have been invaluable, shaping my thinking and reminding me daily of the real-world importance of this work. And to my fellow professors, your support, dialogue, and camaraderie have made this academic journey both intellectually rich and personally meaningful.

I am also thankful to the EXAI Manifesto collaborators, whose commitment to ethical excellence in artificial intelligence helped shape the philosophical foundation of this book. Your contributions to building responsible, human-centered hybrid systems continue to resonate throughout their pages.

A heartfelt thank you to Qi (Elaine) Zhang for her invaluable contributions to the user experience research and early design strategy that informed Chapters 3 and 4. Her pioneering work on the Humainority initiatives has been a powerful example of translating AI into equitable, human-centered tools. I also want to recognize the entire Humainority team – your creativity, dedication, and belief in AI for equity continue to inspire what's possible.

Finally, my sincere thanks to the team at Taylor & Francis for their expert guidance, production support, and stewardship throughout the publishing process. Your professionalism and dedication behind the scenes have not only helped bring this vision to life with care and precision but also added significant value to the final product.

Glossary

A

Accountability – EXAI framework, lifecycle ethics, role-based governance
Adaptability – cognitive, cultural, system-level responsiveness
Agentic AI – autonomous systems with embedded ethical hesitation
Algorithmic Auditing – continuous oversight of AI systems
Architecture of Intelligence – modular, orchestrated design of hybrid systems
Augmented Intelligence – human-AI collaboration over replacement

B

Bias – statistical bias, cognitive bias, AI fairness
Bill of Hybrid Rights – foundational rights for human-AI coexistence
Black Box Systems – opacity in AI decisions and explainability gaps
Boyd, John – creator of the OODA loop

C

Causal Inference – Pearl's work, hybrid reasoning
Cognitive Augmentation – AI-enhanced decision-making and interpretation
Cognitive Diversity – varied human perspectives in teams and systems
Companion Systems – AI designed for co-regulation and emotional alignment
Context Engines – systems that prioritize situation-aware reasoning
Cultural Intelligence – cross-cultural adaptation in hybrid systems
Cyber-Physical Systems – hybrid integration of AI, data, and physical infrastructure

D

Data Ethics – privacy, consent, and ethical data sourcing
Data Visualization – clarity, insight, storytelling through visuals
Decision Loops – OODA, OODA-F, iterative models
Design Thinking – user empathy, problem reframing
Digital Presence – human anchoring in digital interactions

E

ECTM – Experimentum Crucis Technology Matrix
Emotional Resonance in AI – synthetic empathy, user perception
Ethical Auditing – embedded ethics and transparency
Ethical Fallback Loops – system pause/reversal in high-stakes decisions
Ethical Hesitation – intentional delay in AI response for human safety
ETHOS™ – five pillars of (Empathy, Truth, Humility, Ownership, Stewardship)
EXAI Manifesto™ – Ethical Excellence in Artificial Intelligence framework
Explainability (XAI) – core EXAI principle for trust and transparency

F

Feedback Systems – participatory design, auditability
Foresight Models – strategic anticipation of tech impact
Friction in Automation – preserving reflection in streamlined systems

G

Governance Stack – role-tiered hybrid oversight architecture
Ground Truth – model validation and training data integrity
Guiding Principles – from EXAI, UNESCO, EU AI Act

H

HAM – Hybrid Audit Model for real-time, tiered audit
Human ETHOS™ – individual-level ethical compass, coined by **Ng**
Human Perspective – subjective, emotional, and interpretive lens
Human-Centered Systems – design aligned with user values and workflows
Human-In-The-Loop – moral safeguard ensuring human agency in automated systems
Hybrid Architect – designers of human-machine collaboration
Hybrid Intelligence – orchestration between human and AI reasoning
Hybrid Mind™ – model of human-AI convergence, coined by **Ng**
Hybrid Rights – explainability, refusal, presence

I

Imagination in Design – cultural futures, narrative co-creation
Interpretability – transparency in AI outputs
Iterative Design – adaptive, feedback-driven system development

J

Justice in AI – alignment with fairness, accountability, and non-discrimination

K

Kant, Immanuel – origin of the categorical imperative used in EXAI ethical grounding
Karma – reframed as systemic ethical feedback, see also: Human ETHOS
Knowledge Graphs – structured representations that enable explainable inference

L

Lexical Intelligence – language modeling, modular symbolic reasoning
Lifecycle Accountability – ethical oversight from inception to deployment
Local vs Global Maximum – dual-frame ethical design strategy

M

Manifesto, EXAI – company-level AI governance framework (Ng), see also: Human ETHOS
Meaning-Making – human interpretive contribution in hybrid systems
Modular AI – composable architecture, explainability, specialization
Multi-Modular Intelligence (MMI) – framework for hybrid cognition
Mutual Trust Loops – co-adaptive learning between human and machine

N

Narrative Pluralism – respecting multiple perspectives in training and design
Neuro-symbolic Systems – AI approaches combining logic with pattern recognition
NIST Framework – U.S. government's AI Risk Management model

O

OODA Loop / OODA-F – decision-making frameworks with ethical feedback
Orchestration Layer – system logic guiding modular coordination
Oversight by Design – embedding ethical scaffolding into infrastructure

P

Participatory Feedback – inclusion of users in system refinement
Personalization vs. Pluralism – balance in AI-driven user experience
Presence in Hybrid Systems – maintaining human role in automation
Proactive Governance – preemptive ethical regulation

Q

Qualitative Feedback – integrating human nuance into AI calibration
Quantification Bias – over-reliance on measurable inputs in decision systems

R

Reflexivity in AI Systems – system response to social/cultural cues
Responsible Innovation – foresight and governance of exponential tech
Rhythmic Governance – temporal triggers for system checks
Risk Scaffolding – layered approaches to risk mitigation

S

Scenario-Based Ethics – contextualized ethical modeling
Statistical Storytelling – data interpretation through narrative
Symbolic Reasoning – logic and modularity in AI cognition
Synthetic Syndrome™ – coined by **Ng**, emotional impact of invisible AI displacement

T

Technological Inflection Points – moments of irreversible adoption
Translator (Interlude) – cross-cultural design and meaning systems
Trust and Explainability – foundational to responsible hybrid systems

U

Universal Audit Layers – reusable audit architectures across domains
User-Centered Design – systems shaped by end-user needs and perspectives

V

Value Alignment – ethical harmony between human and AI goals
Visualization Principles – data clarity, insight design, cognitive load

W

Whorfian Hypothesis – language's impact on thought and perception
Workforce Augmentation – AI as collaborator, not replacer

X

X-Factor – unpredictable human insight that enhances hybrid reasoning

Y

Yielding Systems – hybrid models that accommodate human override and learning
Youth and AI – implications of AI education and system design for younger generations

Z

Zero-Knowledge Ethics – theoretical future of trustless AI protocols
Zhang et al. – advocates for Composition of Experts (modular AI)

Bibliography

Aafjes-van Doorn, K., Bowers, J., & Przeworski, A. (2022). Pragmatic Augmentation of Mental Health Therapy with Digital Technology. *Journal of Technology in Behavioral Science*. https://doi.org/10.1007/s41347-022-00264-y

Adichie, C. N. (2009). The danger of a single story [TED Talk].

Bousquette, I. (2025, April 18). Johnson & Johnson pivots its AI strategy. *The Wall Street Journal*. www.wsj.com/articles/johnson-johnson-pivots-its-ai-strategy-a9d0631f?refl ink=desktopwebshare_permalink

Boyd, J. (1986). *The Essence of Winning and Losing*. [Unpublished briefing slides].

Bricklin, D. (2009). A spreadsheet way of knowledge. *IEEE Annals of the History of Computing*, 31(3), 20–31.

Bughin, J., Seong, J., Manyika, J., Chui, M., & Joshi, R. (2019). Notes from the AI frontier: Modeling the impact of AI on the world economy. McKinsey Global Institute.

Campbell, J. (1949). *The Hero with a Thousand Faces*. Princeton University Press.

Campbell-Kelly, M. (2007). *From Airline Reservations to Sonic the Hedgehog: A History of the Software Industry*. MIT Press.

Chomsky, N. (1965). *Aspects of the Theory of Syntax*. MIT Press.

Christian, B. (2020). *The Alignment Problem: Machine Learning and Human Values*. New York: W. W. Norton & Company.

Chui, M., Manyika, J., & Miremadi, M. (2018). The executive guide to AI. McKinsey & Company.

Cusumano, M. A., & Selby, R. W. (1995). *Microsoft Secrets*. Free Press.

Deloitte. (2023). State of AI in the enterprise: 5th edition. Deloitte Insights.

Dishaw, M. T., & Strong, D. M. (1999). Extending the technology acceptance model with task–technology fit constructs. *Information & Management*, 36(1), 9–21. https://doi.org/10.1016/S0378-7206(98)00101-3 Retrieved from https://doi.org/10.1016/S0378-7206(98)00101-3

Dumitru, C. (2013). Crucial instances and crucial experiments in Bacon, Boyle, and Hooke. *Society and Politics*, 7(1), 45–53.

European Commission. (2019). Ethics guidelines for trustworthy AI.

European Commission. (2021). AI Act Proposal.

Fei-Fei Li. (2021). Human-centered AI: Building trust with empathy and ethics. Stanford HAI Lecture.

Few, S. (2009). *Now You See It: Simple Visualization Techniques for Quantitative Analysis*. Analytics Press.

Floridi, L., & Cowls, J. (2019). A unified framework of five principles for AI in society. *Nature Machine Intelligence*, 1(1), 65–67.

Fortino, A., & Nayak, A. (2010). An architecture for applying social networking to business. *2010 IEEE Long Island Systems, Applications and Technology Conference (LISAT)*.

Gadre, A., Mialon, G., Altman, R., et al. (2022). MRKL systems: Modular reasoning, knowledge and language. Stanford HAI.

Gardner, H. (2007). *Five Minds for the Future*. Harvard Business School Press.

Gartner. (2024). Why AI pilot projects fail. Gartner Research.

Goodhue, D. L., & Thompson, R. L. (1995). Task-technology fit and individual performance. *MIS Quarterly*, 19(2), 213–236. https://doi.org/10.2307/249689 Retrieved from https://doi.org/10.2307/249689

Harari, Y. N. (2017). *Homo Deus: A Brief History of Tomorrow*. Harper.

Hofstede, G. (2010). *Cultures and Organizations: Software of the Mind*. McGraw-Hill.

IDEO.org. (2015). The field guide to human-centered design. IDEO.

Institute of Internal Auditors. (2021). AI internal audit guidebook.

Jain, S., Raju R., Li B., Csaki, Z., Li, J., Liang, K., Feng, G., Thakkar, U., Sampat, A., Prabhakar, R., and Jairath, S. (2024). *Composition of Experts: A Modular Compound AI System Leveraging Large Language Models. arXiv preprint*. https://arxiv.org/abs/2412.01868

Kahneman, D. (2011). *Thinking, Fast and Slow*. Farrar, Straus and Giroux.

Kant, I. (1785). *Groundwork of the Metaphysics of Morals*. Riga: Johann Friedrich Hartknoch.

Kelly, K. (2005). What Technology Wants. 268. Viking.

Kuang, C., & Fabricant, R. (2019). *User Friendly*. MCD Books.

Lee, S.-G. (2013). The Impact of Cultural Differences on Technology Adoption. *Journal of World Business*, 48(1), 20–29. https://doi.org/10.1016/j.jwb.2012.06.003

Le Guin, U. K. (1986). The Carrier Bag Theory of Fiction. In *Women of Vision: Essays by Women Writing Science Fiction*, edited by Denise DuPont, 43–52. Harcourt Brace Jovanovich.

Marcus, G., & Davis, E. (2019). *Rebooting AI*. Pantheon.

McGilchrist, I. (2010). *The Master and his Emissary*. Yale University Press.

Minsky, M. (1986). *The Society of Mind*. Simon & Schuster.

Nadella, S. (2017). *Hit Refresh*. HarperBusiness.

National Institute of Standards and Technology (NIST). (2023). AI Risk Management Framework (AI RMF 1.0).

Ng, J. X. (2024). Data Insights: Core Principles of Statistical Analysis for Storytelling. SAGE Ensights LLC. https://a.co/d/edWAFvL Retrieved from https://a.co/d/edWAFvL

Ng, J. X. (2024). Mastering IT Innovation: 90-Day Proof of Value Blueprint. SAGE Ensights LLC. https://a.co/d/07ZO4zW Retrieved from https://a.co/d/07ZO4zW

Ng, J. X. (2025). EXAI Manifesto: Ethical eXcellence in Artificial Intelligence (EXAI) Manifesto. Humainority.org. https://Humainority.org/exai-manifesto/ Retrieved from https://Humainority.org/exai-manifesto/

Ng, J. X., & Fortino, A. (2025). Strategic Innovation Governance: Identifying Crucial Use Cases via Experimentum Crucis Technology Matrix (ECTM). www.researchgate.net/publication/395205753

O'Neil, C. (2016). *Weapons of Math Destruction*. Crown Publishing.

Pearl, J., & Mackenzie, D. (2018). *The Book of Why*. Basic Books.

Philip B., Pak-Hang W., Bernd Carsten S., et al., "Ethics by Design for Artificial Intelligence," *AI and Ethics,* 4(2), 265–280, https://doi.org/10.1007/s43681-023-00330-4

Pinker, S. (1994). *The Language Instinct*. William Morrow.

Power, D. J. (2004). A brief history of spreadsheets. DSSResources.com.

Rogers, E. M. (2003). *Diffusion of Innovations* (5th ed.). Free Press.

Sackett, D. L., Straus, S. E., Richardson, W. S., Rosenberg, W., & Haynes, R. B. (2000). *Evidence-Based Medicine*. Churchill Livingstone.

Sapir, E., & Whorf, B. L. (1956). *Language, Thought and Reality*. MIT Press.

Shukla, A. (2025). *From AI Agents to Agentic Intelligence: A Comparative Study of Autonomy, Adaptation, and Ethical Design*. SSRN. https://papers.ssrn.com/sol3/Delivery.cfm/5341 168.pdf?abstractid=5341168

Simon, H. A. (1996). *The Sciences of the Artificial* (3rd ed.). MIT Press.

Spiegelhalter, D. (2019). *The Art of Statistics*. Pelican.

TIME Magazine. (2024). Sougwen Chung's robotic improvisation.

Toronto Declaration. (2018). Protecting human rights in machine learning systems. Amnesty International & Access Now.

Tufte, E. R. (2001). *The Visual Display of Quantitative Information* (2nd ed.). Graphics Press.

Tukey, J. W. (1977). *Exploratory Data Analysis*. Addison-Wesley.

Turkle, S. (2015). *Reclaiming Conversation: The Power of Talk in a Digital Age*. Penguin Press.

UNESCO. (2021). Recommendation on the ethics of artificial intelligence.

Vallor, S. (2016). *Technology and the Virtues*. Oxford University Press.

Venkatesh, V., Morris, M. G., Davis, G. B., & Davis, F. D. (2003). User acceptance of information technology. *MIS Quarterly*, 27(3), 425–478. Retrieved from https://doi.org/10.2307/30036540

von Hippel, E. (1988). *The Sources of Innovation*. Oxford University Press. Retrieved from https://ssrn.com/abstract=2877276

Voss, P. (2005–2023). Cognitive AI research papers and notes. Adaptive AI Inc.

Wikipedia. (2024). CyArk and digital heritage preservation.

Index

For Product Safety Concerns and Information please contact our EU
representative GPSR@taylorandfrancis.com
Taylor & Francis Verlag GmbH, Kaufingerstraße 24, 80331 München, Germany